D0742332

SUBDUING THE COSMOS

Subduing the Cosmos

CYBERNETICS AND MAN'S FUTURE

Kenneth Vaux

JOHN KNOX PRESS
Richmond, Virginia

Unless otherwise indicated, Scripture quotations are from the *Revised Standard Version of the Bible,* copyrighted 1946 and 1952.

International Standard Book Number: 0-8042-0856-5
Library of Congress Catalog Card Number: 77-107325

Printed in the United States of America

Preface

The present book began as a doctoral thesis at the faculty of Systematic Theology and Social Ethics at The University of Hamburg, Germany. Professor Helmut Thielicke directed the research. The author's deep gratitude belongs to this man who so beautifully personifies the theological humanism that is the ethical context of this book. The interaction of disciplined theological insight with complex secular concerns, in this case, the value concerns of medicine and technology, has been pioneered by Thielicke. The loving support of my wife, Sara Anson Vaux; sons, Keith and Bert; and parents, the Bert Ansons and H. Kenneth Vauxes, provided the undergirding that made this work possible. In the preparation of the manuscript, I am indebted to colleagues Albert Moraczewski and Thompson Shannon of the Institute of Religion and secretaries Mrs. Joyce Causey, Mrs. Leila Regan, Miss Jane Schwerdtfeger, and Mrs. Jean Giesberg.

KENNETH VAUX

Contents

Foreword by Colonel Edwin E. Aldrin, Jr.

NASA ASTRONAUT

When Christopher Columbus set sail across the Atlantic, he knowingly embarked upon a great adventure. He commenced his journey with an awareness of the possible dangers inherent in such a voyage and with an uncertainty as to its outcome. This spirit of adventure, which beckoned Columbus and men of purpose throughout history to seek new worlds and new discoveries, spurred man's commitment to the exploration of space. The human spirit will always have this compulsion to investigate the unknown, to comprehend its mystery, and to bring it within the control of human understanding and activity.

Man ventures into space realizing his deepest human potential but responding also to a divine impulse. His total activity is an expression of personal faith that the cosmos is ordered and that man's scientific ingenuity has correctly understood the dynamics of heat and space and motion. In short, he believes that the universe will disclose her secrets to his inquisitive spirit.

This, then, is an adventure that embraces the deepest spiritual and ethical sensitivities of man. It challenges his ultimate convictions and most precious values. It brings to consciousness all the profound questions about the nature of the cosmos, the future of world society, and the destiny of man.

Man conducts a continuing search for the answers to these questions as he calls upon and uses his technology in electronics, systems engineering, and a wide range of other scientific and technological disciplines to extend his functions and achieve his goals. The spiritual and value questions are thus related to the interface of man and his technology, and I commend this book as an attempt to discover some of the deeper implications of man's undertaking to subdue the cosmos.

10 April 1970

I

Introduction and Delineation of Problem

"Tonight the heavens have become man's world."[1] With these words President Nixon marked the significance of man's first step on another world and the significance of a day, proclaimed by one, as "man's puberty, his Bar Mitzvah and Confirmation—the beginning of adulthood of the human race."[2] Without electric technology the Apollo program and all subsequent exploration in the universe would be unthinkable. Man could not subdue the universe without the development of computer science and communication technology, those endeavors that fall within the penumbra of activity we call cybernetics. In the late twentieth century man has brought the world more under the control of his energy in his creation of the electronic global village. The promise of the centuries to come is man's subjugation of extraterrestrial environment to his control and communication.

Capacity entails responsibility. Man's subjugation of environment, both on earth and beyond, carries choice options and value considerations. Subduing the environment is certainly human capacity if not man's commission. Considered theologically it is creation mandate. Man is meant to do this. When man takes hold of the cosmos and shapes it to his will he is engaged not only in technology but in theological activity as well. This book is a reflection on the ethical significance of electric technology and man's use of that power to control environment. This capacity is most clearly called cybernation. Although the concept of cybernation refers to a formal principle of feedback and control (there can be cyclic feedback dynamics in water, air, and society as well as electronics), the concept here is restricted mainly to the electric technology. The working definition of cybernetics in this book is the total theoretical and practical process of changing environment through control and communication utilizing the feedback principle.

This work is an attempt to show a way that ethical responsibility within cybernation can take shape. The unique environment of contemporary man is the cybernated era. His life is situated in the new context that the electronic age brings. His decision-making and relatedness take place in this process. For the Christian this challenge necessitates a fresh interpretation of responsibility.

The structure of this book will be as follows: First we must define and understand the process of cybernation. It must be seen as part of the process of secularization from which it emerges. We must then analyze the humanizing and dehumanizing potential which inhere in the process. With this groundwork done, we can examine two themes of Christian theology which provide source content for the understanding of responsibility. The doctrine of creation, in its anthropological aspect of man as the co-worker with God in subduing the earth, is one source. The exploration of the second theme, that of eschatology and cybernation, will enrich our understanding of responsibility. A final chapter on work and leisure will show the concrete direction that responsible life can take given this understanding. The underlying theme of the work is to establish criteria of responsiblity as these are formed in the life situation shaped by cybernation.

Robert Theobald, the technologist/economist has said, "It may well be that the current analyses of the cybernation revolution should have little to do with the 'dismal science' and everything to do with anthropology, philosophy and theology."[3]

In his last monograph, Norbert Wiener, the mathematician who labeled the second industrial revolution with the term "cybernetics," poignantly calls to the religious discipline to engage this new "Golem" in theological and ethical dialogue.[4] The urgency of this overture carries responsibility that the Christian ethicist cannot ignore. Therefore, this response is an attempt to explore these certain points—to use Wiener's phrase, "where cybernetics impinge upon religion."[5]

The cybernation process colors our theological-ethical understanding of reality and conversely our spiritual-ethical knowledge and valuation determines our perception and use of cybernation. As Pascal has implied, spiritual perception is the cutting edge, determinative of the direction of scientific development. Faith and science are naturally intricated. When man hangs a spaceship in the heavens, relying on computerized contact with and control of that venture, he acts in faith. He believes the universe to be vulnerable to that kind of control and he is right. But nature also rewards that faith. As his technology strikes more harmonious chords with the natural order his comprehension and exploration in the cosmos is deepened. This study assumes that spiritual-ethical and scientific technologic discernment of reality are complementary and indeed synergetic.

The term "cybernetics" is derived from the Greek

word ὁ κυβερνήτης meaning "the helmsman" or "the steersman." Homer uses it in *The Odyssey* in referring to nautical activity, as do Pindar and Aristophanes. Plato in the *Politics* and *The Republic*[6] uses it with reference to the prudential aspect of the art of government. In 1843 the French physicist A. M. Ampère first suggested the word in the modern mode of usage. In his *Essay on the Philosophy of Science*[7] he used the term "cybernetique" with reference to the science of civil government. The Latin term "gubernator" is derived from the Greek; hence, the English word "governor."

The mechanical and electrical usage of the word "governor" also underlies the current meaning of the term "cybernetics." In 1868 J. C. Maxwell, studying the steam engine, theorized on the phenomenon of the governor[8]; the small spinning ball that determined energy input by regulating a valve in proportion to the centrifugal force under which it turned. Maxwell analyzed this control phenomenon mathematically in terms of a feedback loop through which machine output and input are linked so that they can be regulated or programmed.

A corresponding development in the biological sciences at the end of the nineteenth century formulated the concept of "homeostasis." Physiologists saw the way the human organism acts to restore internal equilibrium. The turn of the century saw men like Sherrington, McCulloch, and Rosenblueth discerning neurophysiological phenomena in terms of feedback. With the advent of the Second World War in Europe servomechanisms were adapting the self-regulating system to military purposes such as gun laying.

As early as 1942 sensitive scientists were beginning interdisciplinary study on what Norbert Wiener has called "control and communication in the animal and machine."[9] This new approach to behavior, which claimed broad application across the wide spectrum of agents from animal to machine, became the base theory of modern technology. If the first industrial revolution saw the transition of men to positions of control functionaries rather than human energy sources, this second industrial revolution placed control in the hands of servomechanisms and reduced man to programming function. Cybernation[10] thus achieved the meshing of automation and cybernetic devices into one sweeping analysis and process of production and behavior. The hammer removed man one step from productive manipulation of matter; yet man was the source of energy and control. Automation substituted mechanical processes for human muscle and dexterity; now cybernetics was to substitute electronic circuits for the mental skills and decision-making processes completing the estrangement of *homofaber* from the productive process. In this high order penetration of reality the universe becomes vulnerable to man's control in heightened ways.

In the United States and Europe cybernetics has had restricted definition and application. It has been contingent upon systems viable to mathematical formalization and manipulation such as computer engineering, bionics and control systems of engineering. The space program is a sophisticated example of a systems engineering application of cybernetics. In Russia, however, cybernetics is defined quite broadly as "the new science of purposeful and optimal control over complicated processes and operations which take place in living nature, in human society, and in industry."[11]

By definition, then, we note that cybernetics is the science of control in machine and animal under the guidance of information or feedback. It is the feedback principle that renders cybernetic technology radically different from the older machine technology. In cybernation the given input is continually monitored and any discrepancy between the final objective and the current activity is corrected. When it is claimed that computers can think one refers to this ability to learn and correct by experience. Society can be modified in terms of feedback analysis of certain factors, i.e., economic or political. This self-correcting capacity of the cybernetic machines without continuing scrutiny of the human operator constitutes a part of the ethical challenge. Using the original image, the helmsman is the one who guides behavior (of the ship in this case) in three steps. First, he notes any discrepancy between the actual and the desired course. Then he calculates the energy and form of action necessary to remove this difference. Finally, he chooses a helm position and energy input which will appropriate this correction. Indication, calculation, and selection, or simply control, provide new direction and new indication, thus necessitating new calculation and selection. We are thus introduced to the "feedback loop" which is a feature of all self-regulating mechanisms. Cybernetics, building its behavior analysis on this assumption, coupling this with the assumption that, in principle at least, this process can be mechanized, constitutes the ethical challenge. Wiener states it this way: "Knowledge is inextricably intertwined with communication, power with control, and the evaluation of human purposes with ethics and the whole normative side of religion."[12]

Having delineated theoretical definition of cybernetics, we must now broaden our understanding of the process by exploring its development and current applications. Through the advent of the digital computer and other cybernetic devices, inroads have been made into nearly every facet of man's life. The artifacts run the gamut from simple automatic producers of material objects to the sophisticated analytical and interpretive devices that process data.

Cybernated systems have assumed a variety of tasks: rolling steel, teaching spelling, weaving cloth, mining coal, diagnosing illness—all the while detecting and correcting errors in their own performance. The fearsome anticipations of H. G. Wells and Aldous Huxley are now friendly associates in the productive process.

It is the contention of this book that the most challenging developments for the field of ethics to come from the emerging technological era are to be found at the points where our technological and cybernetic apparatus seeks to duplicate or extend human functions.

Human functions from the most menial to the most highly sophisticated are now assumed by cybernetic devices. Although the fictional anticipations of the authors mentioned above are far from realistic in terms of actual possibilities, the object of their concern is very pointed. It is certainly true that man is now capable of creating superstructures which will loom over him in such powerful ways that control shall often elude his grasp.

All of the functions of the human personality are theoretically viable to duplication. The manual functions were reduplicated early in the technological progress of man. The wheel and the lever appeared early in human history reduplicating the burden-bearing functions of the back and shoulders and the leverage functions of the human joints. Today menial tasks are undertaken without notice by all kinds of electronic devices. Not only is manual extension a possibility of our cybernetic devices but they are often called on to supplement or strengthen the human function. Systems are now being devised to duplicate limb and organ functions. The development of prostheses through cybernetic means is so prevalent that thousands of "cyborgs" now enjoy such increased facility.[13]

The reduplication of organ functions through cybernetic device is a rapidly growing science. Recent popular attention to the technique of transplantation of vital organs has overshadowed in the media coverage the development of the cybernetic devices to duplicate organ function. The phenomenon of the heart transplant has eclipsed for the moment the very significant development of the artificial heart and heart assist which will be mentioned later. The recent British lung transplant has obscured the development of the artificial lung systems that are equally sophisticated in both the United States (particularly Tulane University, New Orleans) and Great Britain. Actually the reduplication of lung function has long been a medical possibility. The iron lung is the primitive example of this. The theological-ethical implications of the man-machine symbiosis are surveyed in recent books by D. S. Halacy and Harold Hatt.[14]

It is when we begin to discuss the possibility of duplicating the human functions of thought, analysis, and decision-making that we begin to see the crucial anthropological and ethical significance of cybernation. Every time technological advance enables man to extend his ability functionally it also raises the possibility of replacement and usurpation. Man now has the sophistication to extend and reduplicate nearly every human function. Dr. John Von Neumann, Einstein's assistant at the Institute for Advanced Studies in Princeton, spoke of brain reduplication during the lecture on the MANIAC computer in 1955. He showed how the computer is conceivably more efficient than the human brain. Faster by 1,000 times in impulse reaction capacity; increased memory capacity by a 1,000 factor. Mathematically, of course, Von Neumann is correct, but he gets into trouble when he becomes prophetic in his theorizing and suggests that "from the turning machine we can envision the universal machine in which self-producing and evolutionary characteristics could be built."[15]

When we postulate this kind of evolutionary significance to the artifact we confront the ethical question raised by Arthur C. Clarke. In his book[16] inspired by the film *2001: A Space Odyssey,* he raises several questions. Is man's technologic venture to unexplored worlds postulated by his violence (the apes with jawbone) or ingenuity? Is his technological creation benevolent or malign (Hal)? Is his venture linear and progressive or cyclic and perhaps regressive (the obelisk and transformation of age: baby and astronaut at end)? The man-machine interface is the context of the ethical problem in this understanding.

The question of the expendability of man is raised when the *Scientific American* made this bold assertion: "The brain's superiority rests on the greater complexity of the human nervous system and on the greater efficiency of human memory. But is this an essential difference, or is it only a matter of degree that can be overcome with the progress of technology? . . . there is no conclusive evidence for an essential gap between man and machine. For every human activity we can conceive of a mechanical counterpart."[17]

Here we see the full import of the cybernetic challenge. The fundamental anthropology that underlies this statement is that the uniqueness of man is somehow carried in his functional, particularly his mental, abilities. Both the capacity that gives rise to this attitude and the attitude itself constitute ethical challenge. Although the computer can play checkers, reproduce itself, and even show emotion,[18] Wiener sees in this the greatness and the danger.[19] The computer certainly can evidence creativity and originality which enhance its challenge.

Feedback devices have also invaded the arts. Television dramas have been written by computer manipulation of plot segments; music has been written and performed by the computer; languages have been translated; social relations have been programmed. All of these innovations rise from the field of communication engineering as opposed to power engineering. Here in communications we see the dimensions of the ethical question. In the important book *Understanding Media: The Extensions of Man,* Marshall McLuhan shows how the media are in a very real sense the extension of man. Each new medium that influences us has a permanent altering effect on our psychic environment, ". . . imposing on us a particular pattern of perceiving and thinking that controls us to an extent we scarcely suspect."[20] Involved inextricably in the cybernated ways we extend communicatively into our environment, the personality is altered. ". . . our human senses, of which all media are extensions, are also fixed charges on our personal energies . . . they also configure the awareness and experience of each one of us."[21]

William Cozart of the Ecumenical Institute in Chicago suggests with the conceptual aids of McLuhan's "Electronic Village" and Chardin's "Noosphere" that the whole of human kind is involved in a transforming captivating way with the cybernated communications media which are our functional extensions. "Electronically the world is being not only brought together in terms of our awareness of events simultaneously, but it is being cooled down together, brought together as a participation kind of experiment, so that everyone is forced to do something with his existentialistic lucidity . . ."[22]

Here we see the dimensions of cybernetics that give rise to ethical consideration. Hans Jonas states it: "There is a strong, and it seems, almost irresistible tendency in the human mind to interpret human functions in terms of artifacts that take their place, and artifacts in terms of the replaced human function."[23]

Some of the most thrilling cybernetic breakthroughs occur in the field of bionics and biological technology. In Houston, Texas, for example, the Texas Medical Center is already making extensive use of the computer in diagnostic work, monitoring vital functions and research. With the use of the organ artifact, the ventricular heart bypass by Dr. Michael DeBakey and his cardio-vascular team, staggering medical possibilities begin to appear on the horizon.

A symposium recently attended by the author illustrates the condition of the present and indicates the shape of the future. It was the annual symposium of the Texas Medical Center on "Biomathematics and Computer Science in the Life Sciences." The theme of this year's conference was "The Medical Uses of Man-Machine Systems." A series of technical papers showed the ways that the mathematical model and the electronic device now contribute to medical diagnosis and treatment.

The computer heart monitor is a good example of cybernetic device applied to patient care. Through electrode sensors the patient's EKG is continually recorded and evaluated. Programming this input with desired patterns gives an output which may be oxygen or drug regulation or an alert signal to the attending physician. Doctors using these highly developed techniques tell about the almost mysterious way that the patient and apparatus relate to each other. As man subdues inner space through electric technology numerous ethical issues emerge which focus on the themes of human identity as the forces of freedom and coercion come to bear upon him. In the movement toward outer space the problems are of course more social-ethical in nature.

It is necessary at this point to reflect on the emergence of cybernetics. The development of "the cybernated era" is a direct outgrowth of the secularization process which has long been at work in Western civilization. Within this movement cybernetics can be understood theologically.

The secularization phenomenon can be viewed from many aspects. It could be observed from the perspective of the political phenomenon of the passing of divine-right social organization. It could be viewed philosophically as the transition in the mentality of people with the rise of scientific philosophy. It could be viewed theologically in terms of a cultural death of God. In this section of the study we will view the secularization process in three dimensions showing how these directly instigate and inform what we have called the "cybernation process." We note first the biblical roots of secularization, particularly in the aspect of the disenchantment of nature. Then we must consider the cultural phenomenon of the desacralization of society as this emerges within the renaissance in Western Europe. Finally we will consider the devaluation of man as this occurs with the enlightenment of the rise of modern science.

Max Weber has used the word "disenchantment" to refer to the gradual process by which nature has been freed from its religious overtones.[24] This process had its inception in the Hebrew creation narratives. The primitive world was a world teeming with spirits and demons.

Magic was the key to unlock the various impulses at work in nature. Everything was alive to the primitive as he stood within nature. Primal energies moved untamed through the cosmos. Occasionally these energies focused in persons or objects. The trees, the mountains, wind and rain, the changing of the seasons, all were phenomena caused by the spirits. The fertility gods, the astral deities, the baals, all were primitively conceived as media of spiritual possession and activity.

The Hebrew accounts of creation begin a departure that is not yet fully emergent. Although strongly influenced by other Near Eastern myths of a more animistic and mythological nature, the essential meaning of these narratives is that Yahweh is distinct from creation. In the Babylonian accounts the stellar bodies were semidivine beings; in Genesis these are creations, formed *ex nihilo,* related to God creatively as contingent and dependent. The divine is in no sense commingled or essentially involved in the creation. God is reflected in his works, not borne by his works. This distinction is most clear in the elucidation of this doctrine in Calvin's *Institutes*: ". . . this skillful ordering of the universe is for us a sort of *mirror* in which we can contemplate God, who is otherwise invisible."[25]

The heavenly bodies, the earth, and all therein reflectively express God's power but inherently can claim no divinity nor evoke by right religious worship. Although God is the sustainer and energizer of the natural process, he always remains over against it in distinction.

Nature is then "disenchanted." It is created as an instrumentality to man's well-being. The sun, moon and stars, the animals, and the vegetation are given to sustain man's life. Far from possessing divine power to create and control man's life, the Genesis world is quite neutral. It is there for man to subdue. He can use it responsibly or he can tyrannize it, in turn being tyrannized himself. God calls the creation into being and names its parts. In extending to man the privilege of naming the animals he exhibits the matter-of-fact objectivity with which man is to approach the creation. This objectivity is not to be exploitative or manipulating. Man's task in the light of disenchantment is to utilize and develop with respect, his trust. "The mature secular man neither reverences nor ravages nature. His task is to tend it and make use of it, to assume the responsibility assigned to The Man, Adam."[26]

Disenchantment of nature has been the necessary precondition to the development of science. A cosmos that is uncanny and spirit-filled does not call forth the experimental enterprise.

Our modern technological culture was further forged in the crucible of the Christian faith. Here is the finalization of the disenchantment process. The moon that is an astral deity or the instrument of the malevolence of a deity does not invite exploration. The coincidence of man's landing on the moon and the revival of astrological religion reveals the contrast between the enchanted and disenchanted cosmos-views that still vie for man's faith.

The Hebrew creation faith revealed God distinct from nature; the Incarnation revealed God in his distinction fully, and released redemptive reality into human life. Here one can find numerous expositions. John Wren-Lewis, for example, states that "the technological revolution in Western civilization is the outgrowth of the Christian redemption process."[27] Jesus Christ, in disclosing God as a loving Father, also revealed, in death and resurrection, that the world is free from the terror of the demonic. Here, as another dimension of the process of disenchantment, freedom is given to scientific and technological impulse. Helmut Thielicke notes that it is

> surely a remarkable thing that science and technology have, besides the root that lives in the native soil of Greek culture, another root in the Christian West... The reason is that through Christ the world was stripped of its daemonic character [*entdämonisiert*], so that men lost their fear of it, gained the ability to be calm and objective, and thus were able to look at it from an observer's point of view ... It is true that the benefit of objectivity gained through the redemption continues to survive for a time after secularized man has long cut himself off from the source of this freedom. But when the ultimate consequences of this secularization are drawn, he again abolishes freedom of investigation, for the very simple reason that the unredeemed man is afraid of the truth.[28]

Though the implication that the universe is daemonic for the unredeemed is questionable, it remains true that the freedom to subdue nature is a gift of faith and faith is the freedom to subdue nature. Although much unfortunate anti-Chinese propaganda accompanied the Apollo 11 venture, it remains true that the mainland ignoring of the event has philosophic as well as political meaning. The Orient has an anti-technological spirit in places not touched by the Western, residually Judao-Christian influence.

The dimension of secularization which we are labeling "disenchantment of nature" rests on the distinction of God and nature, founded on the doctrine of the sovereignty of God in creation and redemption. True secularization, which recognizes the divine source and impulse, motivates science and technology and consequently gives impetus to cybernation. Berdyaev writes:

> However paradoxical it may seem, I am convinced that Christianity alone made possible both positive science and technics. As long as man had found himself in communion with nature and had based his life on mythology, he could not raise himself above nature through an act of apprehension by means of the natural sciences or technics. It is impossible for man to build railways, invent the telegraph or telephone, while living in the fear of demons. Thus for man to be able to treat nature like a mechanism, it is necessary for the demonic inspiration of nature and man's communion with it to have died out in the human consciousness.[29]

As the offspring of that scientific technological venture forged in the crucible of Christian humanism, cybernetics too is nurtured in this understanding of creation. The cybernation process very deeply rests on a disenchanted view of creation. Many of the aspects of cybernetics are such high-level penetration of the cosmos that they come frightfully close to assuming personal characteristics. The view of cybernated devices as personality extension,[30] particularly in the case of the computer as the artifact of the brain and extension of the central nervous system, is only possible on the basis of a thoroughly disenchanted view of the creation. In fact, a quite opposite danger is sensed as one reads some of the nearly "apocalyptic" later writings of scientists, such as Einstein, Wiener, and Oppenheimer. Here the implied caution concerns the re-personalization or demonization of our artifacts as the technological monsters of our ingenuity take on proportions of mastery over their creators. That nature is neutral in terms of spiritual or daemonic energy is a starting point on which cybernation process proceeds.

Professor A. Van Leeuwen has characterizied the

socio-cultural change which marks our history in this technological age in terms of the collapse of "ontocratic" societies.[31] Here we see another aspect of the secularization process that contributes to the rise of a "cybernated era": the desacralization of society.

The traditional "ontocratic" pattern of life was an understanding of society which saw reality as a total harmony between the sacred and the secular. The society in this world view is a sacred entity. Social structures take on an immutable character and thus are no longer subject to change or transformation. Van Leeuwen sees part of the historical process that has ushered in this technological age as the collapse of societies based on "ontocratic" world view. Since the orders of society have lost their sacral character in our day they are now flexible. They anticipate and welcome the innovation; man is free to pragmatically shape his world. He is free to adventure into space knowing he is not violating any sacrosanct cosmic structure. The expressions heard during the moon landing that man should not leave this domain or tamper with other domains rests on this former sacralized "ontocratic" concept of the universe.

Richard Shaull of Princeton commented on Van Leeuwen's analysis at the World Council of Churches Church and Society Conference in Geneva: ". . . In this age of rapid social change . . . technology has emerged as the central factor . . . The technological revolution developed in the West as a consequence of a fundamental change in man's understanding of reality and of the social order."[32]

A social fluidity which is certainly a movement within the process of secularization is a conducive force in cybernetic advance. Because society has lost its sacral character, the components of society: science and industry, education, health and welfare, are open to innovation and transformation under the influence of cybernetics.

The desacralization of society is a movement which emerged first in Western culture with the collapse of feudalism and the rise of Renaissance society. The process, with occasional exceptions like Japan, has yet to claim other non-Western societies. The occurrence of desacralization in Western Europe is certainly a precipitating factor in the rise of rationalism, the development of modern science, and the industrial revolutions of succeeding centuries. This development, when seen as a dimension of secularization, is causally related to the emergence of cybernetics.

Desacralization of society ushered in a new sociological ethos. No longer were the structures of society immutable and inflexible. Society was now subject to change and transformation. Society was now admittedly open to control and planning. The political developments of post-Renaissance Europe, the formation of Protestant political experiments such as Calvin's Geneva, the rise of the Nation States and contemporary welfare states, all reflect a new fluid concept of social organization. One could now plan and evaluate the dynamics of social change; not until the emergence of Marxism was there a new social dogmatism based on historic-sociologic determinism.

This kind of social milieu was fertile environment for a cybernetic concept of control and manipulation. Jacques Ellul recognizes this sociological source of technological society. He uses the concept of technique to identify the cybernetic principle of control and communication: ". . . the reality is that man no longer has any means with which to subjugate technique which is not intellectual, or even as some would have it, a spiritual phenomenon. It is above all a sociological phenomenon, and in order to cure or change it, one would have to oppose to it checks and barriers of a sociological nature."[33]

This social phenomenon of secularization has thrust upon man a high degree of responsibility. Inevitability no longer can be the excuse for resignation and failure to respond with initiative in social determination. Man cannot say fatalistically that prevailing orders and structures are sacred, not to be tampered with. In contrast man is forced to acknowledge the conditionedness of his understanding, his mores, and sacral forms. The national and international responsibility of this implication of cybernation is great. The Secretary-General of the United Nations captured the significance of decision-making in the context of social feedback:

> The truth, the central stupendous truth, about the developed countries today, is that they can have . . . in anything but the shortest run . . . the kind of scale of sources they decided to have . . . It is no longer resources that limit decisions. It is the decision that makes the resources. This is the fundamental revolutionary change . . . perhaps the most revolutionary mankind has ever known.[34]

Theologically understood, this process discloses God's purpose in history. Friedrich Gogarten has noted that secularization is the legitimate consequence of the impact of biblical faith on history.[35] Biblically, the process of secularization as a social process is related to the universalization of the servant role of Israel (i.e., Jer. 31) and to the Apostolic release of the gospel to the Gentiles (i.e., Acts 10-11). All earth's peoples and communities are drawn into the redemptive activity of God. Planetary society is a transformable and transforming entity responsive to the creative God of history. Ecclesiastically secularization is related to the disintegration of the Holy

Roman Empire, the dissolution of the *corpus christianum* as a socio-cultural entity.

The development of what Tillich has called "secular autonomy" is the ultimate outcome of biblical faith and the natural consequence of Christian history. This is not to say that this autonomous culture is given sacral character in some new way; rather it is given autonomy that makes genuine responsibility possible. Protestantism cannot, says Tillich, demand a new situation of control by imposing this or that world view which is theonomous. The spirit of a time prevents this. The task of the Christian is not to long for some preceding era when the dialectic between God and man created a more comfortable ethos. He must rather accept the present in its givenness and seek responsible action in his own time. Responsibility scrutinizes each temporal and contingent reality by a more fundamental criterion.[36]

At this point, cybernation, growing out of the secularization process comes under ethical scrutiny. There is no sacred givenness to a particular technologic or cultural development. When the day of man's first step onto another world is claimed as the greatest event in the history of mankind, we have sheer technological presumption. The event is plastic. It can be good and great. It can be shaped positively by subsequential response that is creative. Alternately, it can be the blackest day of human history if it symbolizes the exportation of earth's pettiness and divisiveness or if the accomplishment is used for political or military terror.[37] Each successive mastery of nature is autonomous, a gift of God and not a historic or spiritual inevitability. It is therefore subject to responsible restraint. As Thielicke puts it: "... the created orders are not dispensed from the sphere of responsible existence. The task of Christian ethics must consist exclusively in putting questions to the secular understanding of reality; in demanding responsibility from it."[38]

In introducing the human factor Thielicke locates the most important theological reflection within the theme of cybernation. This dimension is concerned with the devaluation of man in the "cybernated era" as this stems from the process of secularization. Man's discernment of inner and outer space can be interpreted as an indication of his grandeur or his triviality. Cybernation has grown from a historical ethos which has stressed the devaluation of man.

The Copernican revolution dislodged man and his world from the center of the universe and the Galilean-Newtonian revolution removed him from the universe completely. To say that man is like the machine or that he is a creature to be functionally understood is a development that is consequent of the revolution that accompanied the emergence of modern physics. Man in the Newtonian world view became a "puny spectator" of the vast mathematical system. His nature was discernable as was the natural process in terms of uniformity to inviolable principle and law. Man's uniqueness was seen in terms of inherent working out of inexorable laws: a functional vision. E. A. Burtt in his analysis of *The Metaphysical Foundations of Modern Science* shows vividly the way in which this operational view of man and nature emerged.[39] The initial devaluation of man can be traced therefore to the shattering of the medieval anthropocentricity of the universe. Man is not the center of the universe. If not the physical center of things, then are man and his world the valuative foci of reality? Does man stand in the eye of God?

This transition is given philosophic depth in the analysis of Descartes. One winter evening in 1619 he discovered that "the human mind played no part" in the essential working of nature. Nature was mathematically ordered with foundation in the admirable science (*Mirabilis scientia fundamenta*).[40] To Descartes the significance of this philosophical vision implied that extension became the basic reality of nature, motion the essential dynamic of that nature, and mathematics the key to revelation of mystery. In the words of J. H. Randall: ". . . nature became a machine and nothing but a machine, purposes and spiritual significance had alike been banished."[41]

Hobbes, along with Spinoza, completed the devaluation of man within the mechanistic world view by "placing man and his life at the very core of the great machine."[42] It could now be said by Spinoza in his *Ethics*: "I shall consider human actions and desires in exactly the same manner as though I were concerned with lines, planes, and solids."[43]

Gradually but decisively the mechanization of man took place so that the " 'new science' promised the transformation of man himself, along with all of life, into the measurable and manipulable working parts of the great machine."[44] If all the operations of nature could be reduced to the behavior of minute particles of matter, human life and experience could be devalued and resolved into the question of the size, the configuration, the motion, the position and the juxtaposition of these particles.[45]

Biology and psychology, the yet primitive life sciences, were conceived mechanistically. The clue to knowledge was taken from physics and extended even to the disciplines of theology and ethics: "thus from the principles of the secular sciences to the foundations of religious revelation, from metaphysics to matters of taste, from music to morals . . . everything has been discussed . . ."[46] All the sciences and their respective discernments of reality were restricted to the mechanistic definition. With the addition of the Darwinian element a dynamic was introduced, but the mechanistic mode of conception was extended even to man himself. Benz notes that "the application of Darwin's natural science idea of evolution led to . . . a very dramatic devaluation of man. The traditional anthropology of the Church had tended to neglect . . . the purely corporeal aspect of man . . . Now, man was suddenly considered from the biological-corporeal aspect only."[47]

The post-Darwinians carry through the systematic devaluation of man. Nowhere is it clearer than in the work of Thomas Henry Huxley. Defining man as "conscious automata" he maintained that all human function can be reduced to molecular analysis. For Huxley the imperialism of science extended its hegemony to the highest uniqueness of man devaluing all creativity by ". . . the extension of the province of what we call matter and causation, and concomitant banishment from all regions of human thought of what we call spirit and spontaneity."[48] With Darwin and Freud articulating the ultimate implications of this reductionism, the first half of our century reaped the daemonic fruits of their sowing.

It is not until our own day that science has properly recognized its limitations. Only recently have mystery and depth been rediscovered as anthropological facts. Paul Tillich has had profound impact on the development of rediscovered depth in the behavioral and life sciences in both the East and the West.

Men like Hans Jonas[49] and Carl Friedrich Von Weizsäcker[50] have explored the anthropological richness that rises from mature reflection on biology and physics, inestimably correcting and redirecting our former mechanistic analyses. Despite this new direction the mechanistic view lingers. Modern science stands in an intellectual ethos which deeply colors cybernation as the contemporary expression of a bridge between the disciplines. Ludwig Von Bertalanffy, an alert contemporary biologist, summarizes:

> The evolution of science is not a movement in an intellectual vacuum; rather it is both an expression and a driving force of the historical process. We have seen how the mechanistic view projected through all fields of cultural activity. Its basic concepts of strict causality, of the summative and random character of natural events, of the aloofness of the ultimate elements of reality, governed not only physical theory but also the . . . viewpoints of biology, the atomism of classical psychology, and the sociological *bellum omnium contra omnes.* The acceptance of living beings as machines, the domination of the modern world by technology, and the mechanization of mankind are but the extension and application of the mechanistic conception of physics.[51]

There is something liberating about seeing man in his continuity with nature. A certain honesty in self-appraisal —a certain exploratory wonder—is motivated by this humiliation. But if man is reducible, because of his mechanistic viability, to a machine, the door is open to control and manipulate his life with cybernetic devices. The human mind becomes the model for the computer and the computer for the mind. The denaturing of man can be seen two ways theologically. Man does stand in the natural process, he is the naked ape, but is his being comprehended and exhausted by assimilation into nature? His uniqueness as the self-transcending, God-related being is challenged by this reduction. Herein lies the danger which will be developed in the chapter on the dehumanizing potential of cybernation.

On the other hand, this aspect of secularization has had the effect of placing man centrally in the creation. He is not a stranger or an outsider. That man was not the center of the universe was a discovery he needed to make. His createdness, though unique, is firmly rooted in the whole creation. Though distinct in creation and redemption, even here he is part of the whole created reality. Both of these meanings of denaturation—the exposure of the fall and realization of man's groundedness in the world—are important theological understandings that suggest the shape that responsibility must take.

In concluding this introductory chapter, our task has been to show how this cybernation process constitutes an ethical challenge. The challenge, in summary, can be examined in terms of two questions. The first question implies the affirmation we must give to the development. Is man a cybernetic system? The second question carries with it a concern of caution. Where does man cease to be a cybernetic system; and in the light of this, where are the limits?

We must acknowledge that man is a cybernetic system. He is a creature of reaction and response. His psychic and biologic functions can be viewed within this model. His nervous system does function on a feedback cycle. Recent biomathematical and bionic models have shown a high predictability factor. Man indeed becomes the model for the machine because of the predictability of his biological and psychological response. Wiener makes much of this similarity[52] as does the contemporary Roman Catholic scientist Neville Moray, who locates man's uniqueness precisely in this predictability.[53] Man is a cybernetic device precisely because of the fact that he is rationally ordered and causally purposive. When man is rational the universe opens to his understanding and utilization. This was the classic significance of the Logos correspondence of man's reason and cosmic rationality. Ethically and aesthetically speaking, this is the universal "Deep calls to deep" (Ps. 42:7).

This is no longer a frightening awareness—it becomes a feeling that gives confidence. To say that man is a cybernetic device is not to insult him but rather to assert his solidarity with nature. The human brain is marvelous to comprehend in its intricacy; so likewise is the human genius to fashion an artifact after that fashion. Richard Bellmon of the Rand Corporation comments: "only when we understand something of the working of the human brain, of its memory or memories, of its power of perception and recognition, and above all of its power to create new ideas, will we be able to use a digital computer to something like full capacity."[54]

But man is more than a cybernetic device. Harvey Cox's image of "man at the giant switchboard" depicts man in his mastery over environment in the rather beautiful grandeur of that estate, but it falls short of seeing the vital uniqueness of man. The anonymity and distance that man receives in the technological era is liberating, says Cox, because it gives man more latitude of choice. Surely man achieves some semblance of independence as he sits at the switchboard, masterfully in control of the electronic village.[55]

Man, however, has a uniqueness that is not comprehended so much by his new-found exoneration through urbanization, as in his relatedness to God and the other man. Cox moves too quickly from the basic anthropological question which he raises to his discussion of *Gemeinschaft* and *Gesellschaft*.[56] He fails to discuss thoroughly the tantalizing question he raises. Is man genuinely liberated by the anonymity and aloofness that he enjoys in the "secular city" or even in the complex and wonderful way that he is intricated in the electronic village? Surely the genius of man is found in his ability to rise above every new technological utopia and in his refusal to allow his life to be defined by this or that secular ethos.

Man is defined in terms of his freedom. Freedom does not mean that he can manipulate a technological system to his intent; it rather means that he is non-programmable. Man is different from the machine in that his life does not find its uniqueness defined in its causality but rather in its relatedness. Man is a self-determining creature whose uniqueness is not found in his design but in the organic dynamism of his being as he stands under God, over the world, and in solidarity with mankind. His life links him to God, his Creator. His relatedness to God in apprehension and association distinguishes him in his createdness. His alien dignity in the redemption that is in Christ marks him unique among all the creation. His life, though discernible as a cybernetic system, is profoundly different.[57] Man is not a machine, nor is he an automaton. "The major difference between man and the machine is that a machine, once made, is apart from its maker; but life is a part of its creator. Because of this, man has the capacity of being aware of God in a unique and very personal relationship. He does not arrive at this conclusion by a detailed analysis of his protein molecules, endocrine secretions, and brain waves, but having the capacity, he thanks God for the gift of awareness itself, which alone lifts him right out of the realm of machinery."[58]

The fact that man is programmable along with the rest of nature is something that the Christian is free to recognize. That he is socially programmable is one of the fundamental theses on which all theory and practice of social order are founded. The fabric of a society depends on the implementation of the fact that God has endowed men with a variety of capacities and that each is to perform functionally for the welfare of the total social unit, indeed for the whole of human kind. His life is created to be ordered in a vast variety of ways. There are orders that are fundamental to the perpetuation of self-existence (i.e., livelihood) and orders that are necessary for the perpetuation of the corporate race (i.e., the family). Often these strivings run counter to the similar strivings of others so that social order and division of labor are the only means to preserve the human race. It is not this dimension of programming that we refer to with regard to the ethical challenge.

When man seeks to program the constitution of man so that his integrity is challenged, indeed when alteration of his very nature is sought, the ethical danger of cybernation emerges. To explore the dimensions of this challenge in the dialectical relation with which man must live in history remains before us.

Man has perhaps made the heavens his world. Through the countless communications networks that interweave in outer space he has at least put his imprint on that hitherto unconquerable realm. Through numerous electric signals he now reads the descriptive poetry of the solar system. He now listens and records; soon he will probe and construct. The question lingers. Will man's subjugation be harmonious or discordant with the purposes of the One whose glory the heavens proclaim (Ps. 19:1)?

Cybernetics and Humanization

> O, wonder!
> How many goodly creatures are there here!
> How beauteous mankind is! O brave new world,
> That has such people in 't.[1]

Miranda's words in *The Tempest* reflect an intermingling of optimism and pessimism, a realistic feeling as we analyze the possibilities of the cybernated era. The words were borrowed by Aldous Huxley to entitle his foreboding glimpse into the future of this *Brave New World.* The electronic era has been greeted both as salvation and as destruction. This section will examine the humanizing and liberating possibilities within this development. The chapter will move through three stages. First, the theological criterion of humanization will be established. Then, in pursuit of this objective, we will show how it is facilitated by the release of man from the tyranny of matter. Finally, we will note the liberating effect of the functional replacement of man. These latter two phases will be discussed from within the process of cybernetics.

Humanization is a theme that has been treated variously in the history of Christian ethics. This book will assume the definition of Paul Lehmann who explored in depth the concept of "God active in history" in Augustine, and has built on this foundation an understanding of God as the undergirding power and catalytic agent in the midst of historical change. Lehmann develops the idea of God as the "moving strength"[2] at the heart of events, straining into the future, striving toward both historical and human fulfillment. At the center of Lehmann's thought is the political character of God's activity, whereby he moves within the structures of history to create the conditions for the fulfillment of human life.

Human perfection or humanization is located in Christ. The incarnation in Lehmann's understanding becomes "the humanization of God for the sake of the humanization of man."[3] Man's humanness is released Christocentrically into his existence. It is derived, not inherent; given, not achieved. This is the direction of the Pauline rendering of manhood both in Ephesians and Colossians. In Colossians, however, we have another element. The maturation of human manhood is qualified anthropologically; derivative and contingent in the "riches of Christ" (Eph. 3:8), but conditioned by man's striving (Col. 1:28, 29).

Here is the point at which we can understand humanization as a theological and cultural category. Although man's true being, his essential humanness, is located in God, his destiny is shaped in the world; for this is the context of his being. Man has his being in the world (*"in-der-Welt-sein"*: Heidegger). As Thielicke has pointed out, this world is not merely man's environment (*Umwelt*); it is his world (*Welt*).[4]

In man's cybernetic subjugation of the universe the

meaning of world and environment is critical. Man in the older machine technology could control aspects of the environment yet always remain objective to that environment. In cybernation man intricates himself in the environment, and he is changed in the process. Instead of being exterior to his technology he becomes interior to it. The prosthetic or orthotic extension of the human today serves as an example. Ego-function as well as ego-image is transformed because cybernetic man is thus engaged in environment.

The humanization of man, though in source and direction centered in God, is determined and shaped by his life in the world. Therefore, his technology, the way he manipulates and constrains his environment to serve him, constitutes a secondary source of his humanization. Man's perfection (τελείοτης), though supernaturally derived and destined, is defined and qualified historically as it takes shape in the world. It is at this juncture of theology and anthropology that we can talk about the humanizing possibilities in the cybernation process. Humanization is anchored in freedom to utilize the cosmos for human value. Lehmann, writing from the Reformed tradition, recalls Calvin's location of humanization in the freedom of man. "To sum up, we see whither this freedom tends: namely, that we should use God's gifts for the purpose for which he gave them to us, with no scruple of conscience, no trouble of mind."[5] In Calvin, the creation is conceived instrumentally and is plastic in response to the creativity of man. Man need not dread or recoil in guilt for tampering with things. The universe will open its secrets to man and unleash its energies to man's control. As Leonardo cried, "Thou, O God, dost sell us all things at the price of labor."[6]

Humanization is maturity that is rooted in the freedom of the Christian man. All things are lawful for the Christian man. Nothing in the creation is inherently good or bad: all things take their significance instrumentally as they are utilized to pursue God's will. *Humanization* within this context is the *freedom to creatively coerce the universe to man's disposition, guided by responsibility which recognizes God as Lord, fellowman as trust, and cosmos as instrumental to God's glory and man's welfare.*

Technical freedom is an aspect of humanization. The radical meaning of freedom in Paul is that man not only is free from being the slave of nature and cosmic caprice, but is released to constrain it to his will. Bonhoeffer was suggesting this understanding in his aphoristic thoughts on "man come of age." In the prison papers collected by Eberhard Bethge, he speaks of a book he would like to write, the first chapter to deal with ". . . the coming of age of humanity. The goal, to be independent of nature. Nature formerly conquered by spiritual means, with us by technical organization of various kinds. Our immediate environment not nature, as formerly, but organization."[7]

Cybernetics in large part facilitates this dimension of humanization. Within this phenomenon man has taken on new possibilities of wholeness and advancement. The nineteenth-century view of man has deteriorated under the impact of cybernation. Then, man's humanness was defined in terms of his parts—a functional conception that saw his identity located in the great mechanized chain of Being. Now a wholistic anthropology is possible. When Neil Armstrong stepped down on the moon his words were perceptive: "One small step for a man—a giant leap for mankind." The human capacity was plummeted more deeply in that venture, which was facilitated by electric technology and systems engineering. The exhilaration that marked the earth's observance of the event was in part *awe* and in part *hope,* seeing what man can be and do as the cybernetic capacity enhances man's penetration and subjugation of the cosmos.

William Cozart relates the new humanization possibility in this way:

> . . . since 1940 something has appeared in history that has absolutely changed man's way of grasping himself. Because of it our inner life will never be the same again. It happened in technology. It happened in the social world. It was called many things. Some people simply called it the communications revolution or communications engineering. It got all kinds of popular titles, such as data processing or computer technology. But the most significant and clearest word that has been applied to this is the word cybernetics. This is a word heard every day, but I am wondering if it is not a sleeping giant in terms of the next direction of the human self-image.[8]

Cybernation has humanizing import because it facilitates profounder release of man's genius and creativity.

Humanization is related to the dynamics of energy location and control. While the principle of entropy or decreasing energy pervades the physical universe, man seems to move against the stream. While randomness seems to prevail in the indeterminacy of Einstein's universe, biological life, particularly the life of evolutionary man, seems to be emergent, the recipient of vital energy increase and organization. "Man is just becoming aware that he lives in what biologists call today open systems . . . in systems which do not tend toward disorder but rather sweep into themselves energy from outside that system, feed on it the way flame feeds on wood, and maintain their being, their openness as a system, by taking away energy from their surroundings and building it up within themselves."[9] Man is no longer manipulated, he is self-steering; no longer is he steered from the outside. Humanization is then release—disentanglement from the caprice of nature. This cultural fact of cybernation appears to be directly related to Pauline freedom from the tyranny of "things indifferent" (ἀδιάφοροι).

We must note at this point the concept of "hominization" as articulated by Pierre Teilhard de Chardin, as this is related to humanization. Chardin was a paleontologist by profession; a Roman Catholic priest by confession. Although the term "cybernation" gained currency after his work was complete, he was dealing with the same phenomenon. Chardin built his system on the uniqueness of the energy increase in biological life that we have referred to. "Hominization" is related to humanization in that they both emerge within the evolutionary development of man.

Chardin reflected on the emergence of man within the created order. He noted the periodic surging of cosmic energy at various points in the evolution of man. Spatially "the phenomenon of man" appears at the penultimate level of reality. Undergirded by the levels of barysphere, lithosphere, hydrosphere, and atmosphere, man stands at the level of biosphere with a new level, noosphere, rising from his appearance.

On the time scale psychogenesis, the stage of the arising of man's thought, is preceded by the epochs of biogenesis and geogenesis. The period of psychogenesis in recent history is giving rise to the age of noogenesis. These latter developments are clusterings of energy within the human phenomenon as man in his consciousness, drawn by omega point, which is for Chardin the point of teleological perfection, informs and tranforms his environment.

In terms of cybernation this energy ingathering and

release is the essence of humanization. Chardin describes it in this way: "Hominization can be accepted in the first place as the individual and instantaneous leap from instinct to thought, but it is also, in a wider sense, the phyletic spiritualization in human civilization of all the forces contained in the animal world."[10]

Through the organizing capacity of mind, man is enabled to be the focus of the convergence and communication of energy forces in the universe. In cybernetic language he becomes the agent of input, feedback, and information or organization. Here is the point of humanization. Man has arrived at or has been brought to the apex position of the universe. (Whether the impulse is conceived in Aristotelian terms as in Chardin or in Platonic-Hegelian terms as achievement or realization is not important at this point.) He is become the gathering and dispatching agent in the dynamics of cosmic energy. Unification is located in hominization for Chardin. "The coalescence of elements and the coalescence of stems, the spherical geometry of the earth and psychical curvature of the mind harmonizing to counterbalance the individual and the collective forces of dispersion in the world and to impose unification . . . there at last we find the spring and secret of hominization."[11]

The science of cybernetics is the kind of approach Chardin was reaching for when he wrote, "We need and are irresistibly being led to create, by means of and beyond all physics, all biology and all psychology, a science of human energetics."[12]

Here, we conclude, is the locus of humanization, theologically understood. The increasing organization and refinement within the energy dynamics of the universe has released to man new technical and spiritual possibility.

> A certain sort of common sense tells us that with man biological evolution has reached its calling: in reflecting upon itself, life has become stationary. But should we not rather say that it leaps forward? Look at the way in which, as mankind technically patterns its multitudes, the psychic tension within it increases, *pari passu* with the consciousness of time and space and the taste for, and power of, discovery. This great event we accept without surprise. Yet how can one fail to recognize this revealing association of technical mastery over environment and inward spiritual concentration as the work of the same great force . . . the very force which brought us into being.[13]

In cybernation, the I-World relationship moves to a higher point of humanization as man is placed at the energy focus of the world. Chardin's thought poetically locates this new organization and concentration around which energy gathers as the divine enhancing of the human position. Cybernation is the sign word for the period where energy is brought to focus in and radiate from man in feedback loops which he controls. The ambiguity of this kind of energy control must be acknowledged. The more man controls, the more responsibility is heightened. It remains paradoxically true that the more man controls, the more remains outside his control. Opening new dimensions imparts new possibilities and dangers.

Having suggested this fundamental shift in relationship, we can pursue a second humanizing implication of cybernation, which is the final state of the release for man from the tyranny of matter.

Man's life is always conditioned by the press of matter. In primitive history matter was recalcitrant, buffeting his life. The fundamental elements of the physical universe—earth, air, fire, and water—were endowed with power by the pre-scientific mind. These were the energizing forces in life. So true was this that the earliest explorers of nature organized all reality around these elements. Thales (640-546 B.C.) identified water as the basic stuff of the universe. For Anaximenes it was air. In Anaximander (611-547 B.C.) all reality is derived from or informed by the "boundless" (ἄπειρον), still an uncanny naturalistic force reminiscent of Hesiod's Chaos.

The history of philosophy and then science can be seen as an effort to come to terms with the brute power of matter. In the Greek and Roman drama, matter was a force of relentless and fatalistic necessity, continually thwarting the striving of men. In the Persian religion matter became identified with the power of evil. Throughout the history of Christian thought a constant temptation has been to ascribe to matter inherent evil power. Philosophically speaking, it has been the sheer "rock bottomness" of nature, the force of necessity, that has always defined man's life in terms of corruption and finitude.

Man's sophistication has not emancipated him from this tyranny. Bertrand Russell says that, in the world that science presents to our belief, "man is the product of causes which had no pre-vision of the end they were achieving . . . his origin, his growth, his hopes and fears, his loves and beliefs, are but the outcome of accidental collocations of atoms . . . brief and powerless is Man's life; on him and all his race the slow, sure doom falls pitiless and dark. Blind to good and evil, reckless of destruction, omnipotent matter rolls on its relentless way."[14]

The implication clearly drawn is that matter not only limits human creativity with its driving necessity, but it possesses power in and of itself.

Dr. Emmanuel G. Mesthene, Director of the Harvard

University Program on Technology and Society, summarized this way at the Geneva World Council of Churches Conference on Church and Society in 1966: "What men have been saying in all these different ways is that physical nature has seemed to have a structure, almost a will of its own, that has not yielded easily to the designs and purposes of man. It has a brute thereness, a residual, a sort of ultimate existential stage that allowed, but also limited, the play of thought and action."[15]

Man's history can be viewed as a process of release from the coercion and compulsion of matter. He first protected himself; then he began the long process of harnessing the forces of nature. Agriculture, hydroelectricity, weather satellites, wonder drugs, space stations—each take their place in man's growing mastery over nature and release from the tyranny of terrestial matter.

Man's attitude also changed. He began to see that the forces of nature were not inexorable and antagonistic. He came to see in the seventeenth century that matter could be manipulated, that planned experimentation could actually create new possibilities within nature. The ancient tyranny was being broken and man knew it. The far reaches of outer space were no longer foreboding. The universe was his to explore, to constrain and manipulate, to harness its energy for the enhancement of his well-being. Man is now aware that he can derive immense energy from matter (nuclear fission) and indeed that he has the power to change and shape matter to meet his needs.

Cybernation, what Wiener calls the second industrial revolution, carries this emancipation to another plateau. Not only through technology have we achieved a control over matter, in cybernation we have imposed self-control on matter. The humanizing significance of this is that man is now not only released from the tyranny of matter but extricated from the necessity of fabricating matter.

The perfect illustration of this is the fully automated and computerized factory. From the transformation of raw material into finished product, all the way to the fire-detection system, all functions are carried out by the machine without any personnel. The factory has become the symbol of man's alienation or extrication from the tyranny of matter. Some would even go so far as to call every industrial extension of man a mechanic subjugation of matter. Ludvik Askenazy defines the machine as "a piece of stone, the branch of a tree—the first tool in man's hands —like automized factories and cybernetics are really one and the same thing: an idea put into practice, an inspiration, an observation, something which enables man to subjugate nature."[16]

The cybernetic level of man's subjugation of matter releases him from menial task to creative design. Evaluation, collation of data along with the actual tasks of production, can be taken over by cybernated devices.[17]

This is not to say that the autonomy given to the computer renders it alien and dangerous to the man who designed and programmed it. This development is positive in that it gives predictability and regularity to routine processes initiated by the creativity of the human mind. It is not like the great computer in the recent film (*2001: A Space Odyssey*) who becomes angry at the astronaut who has shunned his favor and locks him outside the space ship.

This release from matter should not be misconstrued as automatic liberation and humanization. It merely enables man to redirect his creative energies into new areas.

There is no necessary spiritual dimension in this development. The release from the tyranny of matter is neutral. Man is replaced at the level of menial chore and is freed to creative tasks with new freedom. Speaking of the liberation Rudolf Allers comments on the work of Maritain:

> Jacques Maritain sagely observes that many critics of American technology, for instance, are guilty of "confusing spirituality with an aristocratic contempt for any improvement in material life ... especially the material life of others." He goes on to suggest that the much maligned gadget is not readily purchased by its maligner, but that it does make material life less overwhelming, and emancipates the human being from the servitude of matter in the midst of the chores of everyday life.[18]

The humanizing force is in fact the creation of new possibilities. The emerging freedom of control of things is what is new about our age. The humanizing potential has developed along with technology's child: cybernation. Mesthene says: "Technology, in short, has come of age, not merely as technical capability, but as social phenomenon. We have the power to create new possibilities, and the will to do so. By creating new possibilities, we give ourselves more choices. With more choices we have more opportunities. With more opportunities, we have more freedom, and with more freedom we can be more human."[19]

Matter imposes its frustration in a variety of ways. The ravage of erosive rain, the tragic aspects of nature's cycle, the barriers of time and distance all frustrate human purpose. The Apollo 11 crew, so clearly harnessed to earth through electronic communications, reentered the brute force of earth's atmosphere only to have the vital communications lifeline once again severed, however temporary. Even space man, for three minutes, is at the caprice of matter.

Humanization is here related to decision-making and the action-response cycle which is the result of a new communicative relationship to one's environment. A new interior-relatedness or integrity is given man as he stands over environment in control and communication.

Industrially man is also released from matter. Electronic devices such as the automated machines and computers have reduced man's toil to a programming function. Economically this has enabled man to transform raw material into finished product without the heavy burden of physical toil.

Technology has made possible the adequate provision of food for all men. The possibilities of population control raise on the horizon the hope of a humanity that lives meaningfully rather than at the bare subsistence level. The dehumanizing forces of pain and disability that so long have held sway over men are on the verge of being conquered. Specific cybernated devices such as artificial limbs and organs have released countless persons into a new and full life.

The partial emancipation of man from his work-relationship with matter and the aversion of many aspects of the delimiting and debilitating effects of matter on his well-being are two dimensions of humanization carried in the release from this tyranny. The range of human possibility and freedom inherent in the release constitute the humanization. This humanization has spiritual dimension: Henry Clark of Duke notes the way that advancing technology

> . . . not only saves the muscles and sinews of man from excessive strain, curtailment of activity, or premature collapse; it also grants the time, the

> vigor, and the psychological space needed for a more expansive spiritual life. Birth control, better health, easier work—all these are blessings not only because they free man *from* a natural evil of some kind, but also because they free him *for* richer development and expression of selfhood.[20]

The humanization located can be described as freedom *from* the tyranny of matter and freedom *for* utilization of matter for the enhancing and enriching of human life. Man himself becomes the focus of possibility and responsibility. His release from the tyranny of matter can constitute blessing or curse. He can claim his opportunity as a freedom to pursue his basic purpose in life, which is to glorify God and love his neighbor. This is his option; yet, it is shrouded in ambiguity because of his sin.

Harvey Cox summarizes:

> The once terrifying forces of nature, the thunder clap and the lightning flash, no longer frighten modern man. He has tamed the wild panthers of the natural world and harnessed their energies for his own uses. But man himself is now the cause of terror. His machines and machine-like organizations can do more damage, or bring more health, than all the thunder and lightning of the aeons put together. But we know that the God and Father of Our Lord Jesus Christ is not just the God of nature. He is also the Lord of history, the supreme sovereign of economic and political life. He is now demythologizing the structures of corporate human existence and bringing them under control, just as he once conquered the natural forces. He holds in his hand a future for this technological man far richer and more brilliant than anything we have yet imagined.[21]

This ambiguity will be dealt with in the third chapter on the dehumanizing possibilities of man's cybernetic subduing of the universe.

The concluding section of this chapter will develop yet another source of humanization carried within the cybernation process: the functional replacement of man.

Cybernation as functional replacement has two aspects. Both aspects are bearers of humanization. Cybernetic devices are used to replace the vital functions of the human being and as such they humanize by reason of sustaining life. The second area of functional replacement is where cybernetic devices replace man in manipulating and utilizing the environment.

In the last decade a wide variety of electrical devices has emerged, the purpose of which is to replace, support, or supplement human vital functions. These innovations, although not free from ethical ambiguity,[22] by and large have had a liberating, humanizing impact. There is much evidence today that the future of physiological support is going to be in the hands of mechanical device rather than replacement from the human donor. It is quite clear, in fact, that the transplantation of human organs has just been a temporary measure preceding the advent of sophisticated mechanical replacements which will not only sustain life but enlarge its possible functions.

The cybernetic devices now provide an interminable list: the artificial lung and other respiratory contrivances; the heart bypass and electric pacemakers, along with the yet primitive artificial heart; the heart-lung pump, which replaces the entire circulatory function and is now widely used in surgery. Cybernetic devices have been used to activate and sustain diaphragm function. The kidney machine, as yet awkward and scarce, shows promise of refinement.

We have mentioned the limb artifacts already in use. Electronic sensors are now used to help guide the arms and legs of paraplegics. With the miniaturized computer it is possible to reduplicate arm and leg movements, reacting to feedback for constant correction, providing control so that even a car may be driven. Some are so bold as to predict that in fifty years it will be possible to replace nearly all vital organs with compact artificial organs having built-in electronic control systems.[23]

Speculatively, the future is latent with possibility. The horizons of cell chemistry and related discovery in biochemistry and biophysics raise the possibility of creating cellular structures so that strains of living cells might be made that would perform simple functions. The biological and the bionic engineers now work at the threshold of discovery where staggering ethical questions relating to human functions are being raised.

The editors of *Progressive Architecture* suggest, with concepts similar to McLuhan's, that "by the year 2,000, then, we can anticipate that man's sensibilities will have been extended by the machine, and that his work will be done by cultivated growths [functional aggregates of produced cells]. Such a development represents a qualitative change, the accomplishment of which may require that we submit our present prejudices to the healing machinations of the computer."[24]

The humanizing character of this aspect of functional replacement is simply the allowance to live with full utilization of all the endowed capacities. Theologically speaking a fully functioning man can, in fact, be less than a man if he fails to fulfill his divine nature and destiny; but, he cannot be considered a fully humanized person if he has lost vital functions to the point of incapacity to relate to self, to others, to the world, or to God. Therefore, in this way, as cybernetic devices sustain or duplicate vital human functions, they take on humanizing significance. The spaceship is a cybernetic device of this type. It extends the human function. It is an orthotic device in the same sense as the automobile. It enhances the human functions of mobility, sensation, and communication. The synergism of this man-machine mutuality is surely humanization when contrasted with proposals to breed legless astronauts[25] who could operate more efficiently in the constricted environment that space work demands.

We must devote our attention primarily to the actual rather than the potential situation. To do this we must locate the ways in which cybernation replaces man in the functions of control and communication with his environment. After stating generally the phenomenon with reference to a specific example, we will interpret several aspects of this phenomenon showing the humanizing potential.

Cybernation has wrested from man's control many of the functions related to his environment. From his primeval task of claiming his bread through his toil, to protecting himself and his habitat from the ravages of nature, to the task of transforming raw material into useful items—one by one these functions have been assumed by cybernetic devices.

In former times man was related to his world at all levels in terms of function. Using the agricultural motif, he was hungry; he harvested and ground the grain, prepared his bread, and fed himself, carefully preparing for the future with new planting. The steps of need, utilization of environment, need-fulfillment, and feedback control constituted the functional progression of his entire relationship to the world.

Social complexity and functional specialization constituted one level of departure from this fundamental relationship. Cybernation, the fulfillment of automation, is another level of departure. The primary functional tasks of contemporary man in Western society are administration and programming. Many men, it is true, still function manually with respect to the environment; many function intellectually. But increasingly these manual and intellectual functions are being usurped by electronic devices.

The once-feared robots of science fiction (Čapek: *R. U. R.*) are now at work in many of the factories of the world. Unimation Company in Connecticut is solely in the business of producing robots which can be adapted to take over many of the industrial tasks given to men. Management speaks of the many advantages of the artificial men.[26]

Here, in one specific industry, we see the human functions of need-analysis and fabrication replaced cybernetically by the computer. The human functions that remain are the *personal* communication at the inception and distribution ends of the production process, and a minor programming function. The example is typical because it shows two aspects of the new situation which are pregnant with humanization possibility: the release from fabrication and the increased interpersonal reality.

With this general statement and illustration before us we can isolate three ways that humanization occurs within this cybernetic replacement of man's functional relation to matter and environment. We note, first, a new freedom in decision-making; second, a freedom to new interpersonal richness; and, finally, a release to pursue spiritual destiny.

Donald Michael, in his study *Cybernation: The Silent Conquest,* describes the way that management is freed to reasoning and decision-making tasks by the computers:

> The computers can produce information about what is happening now, as well as continually updated information about what will be the probable consequences of specific decisions based on present and extrapolated circumstances . . . freeing management from petty distractions in these ways which permit more precise and better substantiated decisions, whether they have to do with business strategy, government economic policy, equipment system planning, or military strategy and tactics.[27]

Man's humanization is always related to his freedom of decision. His responsible self is shaped as he confronts his environment with choice options. Directly related to Christian freedom and humanization is man's capacity to make responsible decisions. Within cybernation the complexity and significance of decisions are enlarged. Man must be free in this context to willingly transfer to cybernetic devices those functions of which they are capable. He must also have the courage to assume the new decision tasks afforded by that forfeiture which are highly complex and fraught with ambiguity. Wiener says that man must be willing to "render unto man the things which are man's and unto the computer the things that are the computer's. This would seem the intelligent policy to adopt when we employ men and computers together in common undertakings. It is a policy as far removed from that of the gadget as it is from the man who sees only blasphemy and the degradation of man in the use of any mechanical adjuvants whatever to thoughts."[28]

Man is set free for decision at a higher level each time he is functionally replaced by cybernation. The production manager who does not have to decide how many automobiles he can humanly sweat out of the workers is free to consider the problems of accidents and air pollution, i.e., the safety responsibility in his production. Large corporations freed from the mechanical necessities are increasingly becoming places humming with consideration of social implications of production, dynamics of interpersonal and intercorporation relations, evaluation and decision-making.

Dietrich von Oppen, a German sociologist, sees in this development great humanizing possibility. He interprets the breakdown of restrictive communal bonds and the emergence of free functional organizational relations as a liberation of man for personal decision.[29]

On a larger scale, cybernation raises crucial national and international problems requiring decision. If man is technically capable of thus utilizing the resources of the good earth, he is also responsible to see that this is done for the benefit of mankind. Communications have shrunk the globe so that we are aware of each other's needs. In our theoretical grasp and technological competence are all the means to feed the hungry, shelter the exposed, and heal the sick. The mature social reflection on the feats of the Apollo program has inspired the world to see ways to apply technological sophistication to the array of pressing human problems: those plaguing contradictions that sour the taste of our most outstanding achievement and make our victories Pyrrhic in contrast. If man evades the responsibility that is carried with this new capacity, he will reap judgment. If technology becomes for man an escape into personal indulgence and national affluence at the expense of international welfare, he will have forsaken his trust.

Gabriel Marcel sees in this development a crisis of personal and corporate decision. It has humanizing possibility cautiously dependent on man's response. Commenting on Marcel's works Henri Quefellec notes that Marcel

> . . . has joined in accusing technology of degrading man: he refers to all those methods of achieving comfort and material control over the world which have been so suddenly made available to man, and are now his without demanding from him any real mental effort. What is wrong is not that we have hot and cold taps in the bathroom, but that we take them for granted, as the natural privilege of twentieth-century man, his inalienable advantage over his fellow man who dwelt in caves . . . [30]

The functional release is a humanizing force only when man responsibly uses his new realm of decision. He has not assumed control of history nor has he been given power to build a utopia within society. His new freedom is tenuously balanced by his responsibility as the sovereign God sends each privilege. With each advance the implications of his decisions deepen. Pope Pius XII reminded the technologists of this fact in 1958:

> Your duty in organizing the world is not to build a final terrestrial society but to make easier, here on earth, for yourselves and your contemporaries, the only search and the only discovery which really matter; those of God. The meaning of history does not belong to technology. It works to make that meaning clearer, and we may believe that it will make it even more clear: our salvation does not come from technology, which must always remain a temporary and terrestrial secondary cause.[31]

There is humanizing possibility within the decision context of the cybernated era. This decision-making necessity is a form of the new power that cybernated man experiences over his environment. When this power relation to environment claims man from his former position of subservience or at best competitive struggle with the world, he receives possibilities to be more human. His being, from the existential perspective, is enhanced. Even Karl Jaspers with his strong apprehension regarding the "technicization" of man acknowledges the link between being and power.

> The being [*Sein*] of man grows with the fullness of the organic substantiality of his community, of which his creative personality is the representative; he comes to know himself in his reflection of the community. The power of man grows, however, with his mastery over nature, won by planning, cooperation and through the creation of the technological environment. Still the being and the power [*Vermögen*] are not mutually exclusive.[32]

When Ayn Rand characterized the moon landing with the dictum "What hath man wrought" she celebrated a burst of humanizing power which, considered theologically, is man's taking dominion over "the works of His hands"; thus fulfilling both being and destiny.

At this point of intersection between man's decision-making capacity (technological power) and his being, we find another aspect of humanization, contingent on the functional replacement of man. Man is the recipient of a new level of interpersonal reality within the cybernetic process. The man set free from the functional task can find fulfillment in life both within the interpersonal relations that are possible as man assumes rational control over environment and in the apprehension of the rationality of nature which his vantage point allows him to perceive.

Cybernation, in extricating man from the fabricating process, has depersonalized the machine and repersonalized man. The early industrial process did just the opposite. Machines were able to facilitate production, but only as man surrendered his individuality to the machine. He punched the time clock, manned his machine, and for all practical purposes became a part of that machine for the rest of the day until he punched the time clock and went home. Sometimes he could not unwind, and his whole existence took on the character of an automaton. Personal relations, even those within the family, were often mechanized to the point where communication and openness were lost. Many a man, victim of the first industrial revolution, like John Henry literally "died with a hammer in his hand."[33]

The second industrial revolution released man from the machine. Figuratively and often literally he was unemployed (no longer *folded in* the machine). He now had the sometimes threatening opportunity of relating to people as frequently as things. The I-Thou context of his life usurped the previous I-It context; frequently the transition was difficult.

The humanization possibility lay in the I-Thou context. The I-It relation was posited on the utilization, domination, or control of an object.[34] This was his new relation to production in cybernation. No longer was there intimacy in his relation to the machine. He was free, extricated, autonomous, and independent. He could appreciate persons, not because of the disciplined way they functioned, but because of the control and guidance they could now exhibit.

The humanization becomes very precarious at this point. There is much evidence that the contrary is actually occurring: that man is being depersonalized while the artifact is being personalized. This point will be examined in the third chapter. Suffice it to say that man in his detachment from the productive process is free to grow in integrity and interpersonal respect if he chooses. Man can be human despite the dehumanization of the artifact. He can also be exonerated and remain inhuman. The circumstantial factors are heightened, however, in the cybernetic capacity.

Removed from the functional necessity, man can find humanization as he reflects on the rationality of nature. The world has proven viable to his organizing work. Order has been revealed as he has constrained the environment to meet his needs. The computer would be a hoax were it not for the highly predictable order that informs nature. C. A. Coulson, an English physicist, claims:

> There is something else that the machine has given us. For, as Gilbert Murray once said in a broadcast, "The machine is a great moral educator. If a horse or a donkey won't go, men lose their tempers and beat it; if a machine won't go there's no use beating it. You have to think and try till you find what's wrong." I should not agree with Gilbert Murray if he meant to imply that losing your temper with a horse would help. But I do most certainly agree with him that in its insistence on the significance, the power, the coherence and reliability of thought, the machine is a constant reminder to us that the universe is rational, and in its rationality it tells us something of God.[35]

Albert Einstein was also sensitive to this dimension of reality that scientific perception could impart to the world understanding of man:

> If it is one of the goals of religion to liberate mankind as far as possible from the bondage of egocentric cravings, desires and fears, scientific reasons can aid religion . . . Science seeks to reduce the connections [of rules governing facts] to the smallest possible number of mutually independent con-

> ceptual elements. It is in this striving after the rational unification of the manifold that it encounters its greatest successes . . . In this domain . . . one is moved by profound reverence for the rationality made manifest in existence. By way of the understanding he achieves a far-reaching emancipation from the shackles of personal hopes and desires and thereby attains that humble attitude of mind toward the grandeur of reason incarnate in existence.[36]

Einstein was not able to affirm a personal deity but his perception was true. The profound significance of man's functional release is the humanizing possibility of perceiving the logos ordering of reality (Rom. 1:20) and through this, the logos incarnate, who is the origin, sustainer, and consummation of all created reality (Col. 1:15-20).

The Christian is responsible to show the origin and impulse of this development. Nobel prizewinner physicist Arthur Compton tells of a meeting with the Indian scientist Sir Shanti Shatnaga. In a conversation, recorded by C. A. Coulson, "they were talking of the future for India, and Sir Shanti turned to Professor Compton, saying: 'There is one thing that you in the West can teach us in the East. It is something that matters tremendously. Show us that it is good to live in an industrialized community.' In the deepest level of understanding, only those who have seen the incarnation and know in their lives the great Christian doctrine of creation, are big enough for this job."[37]

This leads us to the concluding aspect of humanization which rises from the functional replacement of man in cybernation. Man can be released to pursue his spiritual destiny.

Man's temporal destiny is to be the co-worker with God in the ongoing process of creation. Making human

function of the *Zeitgeist.* This functional definition of man's place in nature is radically opposed to the Christian view of man. Here man is original, not a function. He must be defined in terms of being a person, not in terms of function.[40]

In other words, is "man" (*humanitas*) merely the code word for a specifically defined functional value within the social structure which would make humanity merely the means to achieve an end, namely, production, or is it independent of all achievements, that is, an end in itself? Historically speaking, the Bolshevist ideology decides in favor of the view that man is the performer of functions, while the so-called "free world" at least in its fundamental program takes the stand that man (*humanitas*) has intrinsic value.[41]

The scientific philosophy of life, along with the rationalistic world view and the industrial reordering of life style, all emphasized the objectification of man. Now with the release that cybernation brings, a renewed emphasis on the subjectification of man is evident. This is humanization; for as David Cairns has pointed out with reference to the thought of Emil Brunner: "Man's nature . . . must be considered as 'object' [physiochemical] but primarily as 'subject,' the possessor of reason, culture and art."[42]

A caution: Man is not automatically humanized by virtue of the fact that he is the child of the cybernated age. Man is still a responsible being. He was not necessarily reduced to a functionaire before his liberation. Man stands before God a free agent to choose to be human or to surrender his humanity. Dietrich von Hildebrand, in contrast to Marcel, has emphasized this:

> Gabriel Marcel has emphasized the deteriorating power of modern social organization which makes out of the human person a mere "functionaire." He

> is nothing but the man who punches tickets, hands out stamps, throws a switch, and so on. This is certainly true, but it is not, it seems to me, the primary determinant. Man is not de-individualized by the role he plays in the social mechanism: he must be de-individualized first to play this role.[43]

Within cybernation man is given the option to stand with maturity in a position of dominion over the creation.

When astronaut Edwin Aldrin quoted from Psalm 8 from Apollo 11, the vivid power of man's station under God and over the world was felt. In his release from the captivity to the slavery of necessity, man is freed to maturity. This is man's humanity: to be the mature son of God, claiming and responsibly undertaking his dominion over the earth. The culture that refuses today to take the risks of maturity is a dehumanizing force in the world.

John Wren-Lewis, an English physicist and theologian, claims that

> . . . a culture which restricts human creativity to the cultivation of the natural world within the limits of a set pattern . . . is just as much motivated by the desire to avoid responsibility for the state of the world as is the frightened neurotic who lives by compulsive private rituals. In fact the moral stability of societies governed by "belief" is not a safeguard of man's humanity, as modern defenders of the traditional human outlook usually claim; it is a stability purchased at the price of inhibiting the expression of man's humanity.[44]

Functional release is humanizing not only because it releases man to higher level function but because it accents his being-awareness. Henry Clark summarizes this point and develops the preceding idea of the functional release of cybernation, showing with reference to Hannah

Arendt the release of energy for action and with Harvey Cox the release for service and self-employment. He concludes by noting the spiritual dimension of the release:

> Freed from the necessity to be employed [literally folded in], in any way not of his own choosing, man will for the first time be in position both to celebrate the beauty and joy of his being, and participate in the wonder of being. Delivered from the exaggerated ego-awareness of the Faustian man, and therefore from many anxieties about what he ought to be doing, man might be able adequately to appreciate the grace that has bestowed upon him the privilege of being. Man might then be able, in short, to glorify God and enjoy Him forever.[45]

For modern technocratic man pursuit of spiritual destiny is at once easier and more difficult. Just as perception and freedom are enhanced so temptation and responsibility are enlarged.

The chapter has explored the humanizing aspects of cybernation. The meaning of humanization as a theological and cultural category has been noted. We have seen this category related to the concept of freedom in Calvin and "hominisation" in Chardin. We have established the humanizing impact of man's release from the tyranny of matter. Finally, the humanizing force of the functional replacement of man in the productive process and basic man-world relation has been examined, noting in conclusion the decision power, interpersonal reality, and spiritual possibility carried with this functional replacement. We have been realistic in raising the ambiguity and tension ever present in this development. The next chapter will locate the dehumanizing possibility within the cybernation process.

Cybernetics and Dehumanization

Carl Sandburg as early as 1910 recognized an ominous future of the mechanized society in his poem "The Hammer." Cybernation likewise presents mankind with an ability that despite all the liberating possibilities is fraught with dehumanizing potential. Science fiction writers like Arthur C. Clarke as well as scientists like Norbert Wiener have located the danger as that which is controlled comes to control the controller. This chapter will attempt to locate this negative dimension. The great ecological crisis—the destruction of environment as man distorts his activity of subduing the cosmos—cannot be dealt with in this book. Although the dehumanization of this tragedy is clear, our emphasis must be on those cybernetic extensions which directly make him less than a man.

The working definition of dehumanization will be initially established. Then the underlying problem of the disparity between man's self-control and control over nature will be developed. Two areas of dehumanization will then be described: first, the problem of slavery and subordination to his functional extensions; second, the impersonalization of cybernated man and his society. Finally note will be taken of the secularizing of God that develops when technical progress is glorified.

Within the context of cybernation, humanization has been defined as the freedom to creatively coerce the world to man's disposition, guided by responsibility. Negating this definition would define dehumanization as slavery to the caprice of man's creation or the surrender of human fullness in the cybernetic process.

Man cannot dehumanize himself in terms of his identity before God. His *imago dei* character renders this impossible. Even in complete degradation the mark of his createdness remains. He can violate and neglect his basic nature but even then he remains the deposed king *("roi déposse de"*: Pascal), related to God even in estrangement.

He can, however, dehumanize himself in terms of depersonalization. Speaking in personalistic rather than ontological terms man can deny his relatedness to fellowman and the world. He can shrink from a responsible posture at any point of his relatedness and render himself something less than a human being. This dimension of his life, his functional life where he relates to his environment in terms of manipulation and production, is the place of potential dehumanization. Man has the ability to depersonalize the other by treating him as an object rather than subject, and the option of depersonalizing himself by treating an object to which he is related as subject.

Theologically the danger can be located within the mentality of idolatry as it is derived from the Old Testament. The fundamental threat in the idolatrous situation emerges when one gives personality or power to an object, transferring obeisance from God. The Old Testament speaks of the perpetual tension in the dynamics of faith, tension arising from the subjectification of an object. The first commandments relate the idolatrous situation to the fabricating work of man's hands. The graven images *(pesel)* are derivative from his handiwork, his carving (פסל). So intense was the iconoclastic impulse of the deuteronomic spirit that the very act of fabrication was dangerous even before man located power in his artifact (Exod. 20:23; 34:17).

A twofold act could constitute idolatry. Man either located supernatural power in the artifact or he attempted to control or define the deity in terms of the artifact. Humanness was distorted in the process. When power was located in the fabricated object, man's humanity was distorted because of the misplaced trust or expectation that he brought to the object. When the transcendent God was conceived of as being confined to the artifact, man's humanity was distorted because of the misdirected ultimate relation.

Paul Tillich describes idolatry in terms of a faith unable to maintain the distinction between subject and object. Failing to distinguish, man raises preliminary objects to ultimacy, and the self is dehumanized through what Tillich calls "existential disappointment." "The more idolatrous the faith the less it is able to overcome the cleavage between subject and object. For that is the difference between true and idolatrous faith. In true faith the ultimate concern is a concern about the truly ultimate; while in idolatrous faith preliminary finite realities are elevated to the rank of ultimacy. The inescapable consequence of idolatrous faith is 'existential disappointment.'"[1]

Space technology again serves as illustration. The systems analysis that led to the first moon landing exhibited the precise way that man's technology served his extremely ambitious objectives. This engineering was nearly perfect instrumentality. It served faithfully even in those precarious moments of re-entry when man put himself as it were at the mercy of the guidance systems. On the other hand, we move close to an idolatrous situation when this technological feat is glorified. "What hath man wrought?" suggested Ayn Rand as slogan for man's first utterance on the moon. "This is the greatest day since the creation of the world," said a jubilant president. For others the feat documented their notion that the cosmos was God-forsaken. The former response was wholesome. It prompted exhilaration and awesome joy as exemplified by the astronauts' allusion to Psalm 8, "what is man that thou art mindful of him?" The latter reaction seems to create only temporary satisfaction followed by bitterness. Humanization can be posited only in a universe with transcendent reference. When man alone is glorified progress loses its joy. Both Rilke and Nietzsche exhibit the pathos of a world view where man must transfigure himself (or his work) into a god, because there are no gods. The attempt only leads to retreat from and disgust with a world that apparently cannot sustain itself.[2] C. P. Snow presents a contrary but related thesis when he comments on the space adventure. He claims that the scientific achievement itself rather than man's glorification in it is a cause of failure of nerve and loss of the creative impulse.[3]

Dehumanization within the cybernated process is the *inability to treat man's fabrications instrumentally as objects or to treat his associates in the productive task as subjects.* Within the cybernated process these dangers are present in such a way as to constitute an ethical challenge.

A primary source of dehumanization arises because of the incapacity of man to control his own life in measure concomitant to his mastery over nature. Because self-control and environment control are in disparate relation man tends to be controlled as much as he controls. Control and communication are the secrets of man's cybernetic relationship to his environment. Wiener has noted that existential control is as necessary as environmental control. "To live effectively is to live with adequate information. Thus, communication and control belong to the essence of man's inner life, even as they belong to his life in society."[4]

The recurring theme in Wiener's ethical work is this: Man has increased the sophistication of his communication relationship to environment but he has not developed the same kind of information sophistication in terms of his interpersonal and international relations. Man's intelligence and the benevolent use of that intelligence hold the key to a promising future. " . . . the new industrial revolution is a two-edged sword. It may be used for the benefit of humanity, but only if humanity survives long enough to enter a period in which such a benefit is possible. It may also be used to destroy humanity, and if it is not used intelligently it can go very far in that direction."[5] Man has a communicative relation to world and fellow humanity. His language is essentially a feedback attribute. He orders a machine or he addresses a person. In both cases a signal goes out and is received, and a signal of compliance comes back. Speech is in fact the distinction of man. Man extends himself into his environment through his communication. ". . . where a man's word goes, and where his power of perception goes, to that point his control and in a sense his physical existence is extended."[6]

Wiener's point is that man's control over self and society has not developed *pari passu* with his control over production. The urgency of man's communicative controlling word has not been expressed as urgently as the situation requires and often it has not been heard, let alone returned.

The fascist state marks a failure of one type. Here men are organized according to individual functions. Each individual pursues in isolation his function, seeing it as a contribution to the collective within which he is only a functionaire. Communication or dialogue is a threat to the fascist state. Here decisions are to be made autonomously with reference only to the objectives of the administration. The dehumanization condemned at this point is most poignant because of Wiener's Jewish humanitarianism. ". . . this aspiration of the fascist for a human state based on the model of the ant results from a profound misapprehension both of the nature of the ant and of the nature of man. . . . Those who would organize us according to permanent individual functions and permanent individual restrictions condemn the human race to move at much less than half-steam."[7]

The collectivist society of the Marxist variety poses the same threat, the difference being the concept of inevitability in history which ameliorates or supresses free communication. If historical change occurs in an inevitable way, coalition, accommodation, and dialogue are irrelevant if not impediments to historical necessity. Within both these socio-political forms, Marxism and Fascism, information is not taken seriously and the human dimension of control diminishes as environmental control develops. Eugene Rosenstock-Hussey, in his study of Western man, has noted the way in which Russian philosophy reconceptionalized man as an energy force: "The gospel preached [in Russia] is that he be changed into a force . . . an element in the electric stream that organizes production."[8] Man became the conductor of production energy. "A man was welcome if he could conduct electric current, new energies. If not, nobody was interested in him."[9]

In a recent memorandum Andrei Sakharov, the father of the Russian hydrogen bomb, also locates this danger with reference to Wiener's thought:

> We also must not forget the very real danger mentioned by Norbert Wiener in his book Cybernetics, namely, the absence in cybernetic machines of stable human norms of behavior. The temporary, unprecedented power, that mankind, or, even worse, a particular group of a divided mankind, may derive from the wise counsels of its future intellectual aides, the artificial "thinking" automata, may become, as Wiener warned, a fatal trap; the counsels may turn out to be incredibly insidious and, instead of pursuing human objectives, may pursue completely abstract problems that had been transformed in an unforeseen manner in the artificial brain . . . Such a danger will become quite real in a few decades if human values, particularly freedom of thought, will not be strengthened, if alienation will not be eliminated.[10]

This notable memorandum from the Soviet Union illustrates the way in which cybernetic technology, manipulated by hostile ideologies, constitutes a real danger to man's future.

Dehumanization is carried in both fascist and collectivist systems because of impersonal relations and increasing subservience to the artifacts of production. These both occur because of lack of administrative control. It is as if man set out to subdue the world, accomplishing this without the adequate ordering of his own life or his society. The transformation of the world had placed possibility in his hands that he could not control.

Gyorgy Kepes of the Massachusetts Institute of Technology has spoken of this from an artistic point of view:

"Where our age falls short is in the harmonizing of our outer and our inner wealth. We lack the depth of feeling and the range of sensibility needed to retain the riches that science and techniques have brought within our grasp. Consequently we lack a model that could guide us to reform our formless world."[11]

Kepes goes on to develop the necessity of communication and interdependence if man is to catch up with his technological capacity. To facilitate genuine communication man must face honestly the information available concerning himself. It is interesting that Einstein's principles of relativity, the profound conceptual tools rendering the universe more viable to man's comprehension, were paralleled the early part of this century by Freud's discoveries of the unconscious motivations of the human mind. It is as if a highly rational external universe opened itself to man's search while at the same time a highly irrational internal universe disclosed itself as the threat to all his striving. Freud spoke of it frequently with the metaphor of an iceberg of uncontrollable motivation beneath the placid surface of man's new-found environmental control.

A biographer of Freud has commented on the significance of his work: "The control that man has secured over nature has far outrun the control over himself . . . man's chief enemy and danger is his own unruly nature and the dark forces pent up within him."[12]

Many scientists have been aware of the perverse motivations of man which can distort the fruits of his technology into destructive capacity. Because of a lack of humanistic control and unworthy international political motivation many great scientists have withheld the results of their research or called for humane implementation of their techniques. After the Second World War Norbert Wiener announced his refusal ever to publish work that "may do damage in the hands of irresponsible militarists."[13] Robert Sinsheimer, noted biophysicist at the California Institute of Technology, recently called on his scientific colleagues to

> ... emerge from their laboratories to exercise their prophetic function—to become responsible prophets to the people ... It is our ability to forecast the consequences of our actions that engenders our responsibility for them ... We in science are growing up now. Our toys become more potent. The little games we play with nature are for great stakes, and their outcome moves the whole social structure. We must accept our responsibility.[14]

Physicist J. Robert Oppenheimer belatedly notes this dimension of responsibility:

> The experience of the war has left us with a legacy of concern. Nowhere is the troubled sense of responsibility more acute ... than among those who participated in the development of atomic energy for military purposes ... the physicist felt a peculiarly intimate responsibility for suggesting, for supporting, and in the end, in large measure for achieving, the realization of atomic weapons. Nor can we forget that these weapons, as they were in fact used, dramatized so mercilessly the inhumanity and evil of modern war. In some sort of crude sense which no vulgarity, no humor, no overstatement can quite extinguish, the physicists have known sin; and this is a knowledge which they cannot lose.[15]

Oppenheimer locates this deep ambiguity in man's technological advance, the disparity between his control over environment and his self-control, and rightly calls it sin. The danger is found in the inability to conceive of one's technical ability as instrumental to the benefit of mankind. Dehumanization results when man's technical creations override his ethical capacity of control and overpower him. Often the problem is centered in false trust in or devotion to the fruits of one's technology. Physicist Ian Barbour is correct when he analyzes the situation and calls our culture at the two points of personalization of things and depersonalization of persons:

> Our culture is increasingly the servant of external, technical interactions of control and manipulation (what Buber calls I-It relations) to the neglect of personal response to people as subjects (I-Thou communication). Persons should be ends, and things means, not vice versa. We are called to love people and use objects, rather than loving objects and using people. The machine tends to set the pace for man, requiring him to adapt his schedule to its needs. Somehow the machine can take possession of a man's life, and the relation of the craftsman to his work is lost.[16]

Cybernation holds dehumanizing potential in measure unknown to the experience of man. Complete mastery over nature is the goal when self-organization of the productive processes is sought. Here man's Promethean defiance is most tempted. No power is so sweet to man as the power to control. Pride can readily take the form of hubris or self-sufficiency so that even the prerogatives of God are claimed. William Pollard, the famed scientist, has written of this danger. He sees within the scientific community the temptation to believe in self-sufficiency. Also evident is the lack of existential control evidenced by hubris instead of humility when mastery over nature is achieved. "No more terrible affront to the creator could be made by man than this: all out determination to seize

God's creation from him and make himself sovereign within it."[17]

It is at this point that our understanding of responsibility must be examined. If man seeks to exert lordship *(Herrschaft)* over the environment, we must acknowledge that this is not necessarily evil. This is the task he is given to in his apex position in the creation of the world. The critical question lies in the purpose he sees as the result of his dominion. If man is motivated to transform the environment in the service of human need and for the enrichment of human life, he is motivated responsibly. If, on the other hand, he seeks to radically alter the environment so that he might subjugate the world and his fellowmen to some perverse purpose, he acts irresponsibly. Most often ethical problems are couched in great subtlety. The good of the wood pulp industry as over against the disaster of deforestation is example, and the discernment of motivation is difficult. Every man and most societies feel that they are motivated by genuine humanism. Hitler and Stalin felt that their systematic control and regimentation of environment and society were for the good not only for the nation but for mankind in general.

The corrective balance comes through international law and justice best achieved when man recognizes that the ultimate sovereignty in the world belongs to God. In this context man can perceive that God wills to share his dominion with and through man. Responsible cybernetic planning then becomes a search for the benevolent will that transcends personal power impulses and seeks the common good. Responsible planning stands under this perspective so that the transforming design of man is humbled by, encouraged with, and guided toward the will of God who seeks to transform all reality. In this *understanding* cybernetic impulse is seen as stewardship.

Man's increasing dominion over the earth must always be seen as a trust for which he has stewardship responsibility. Only in this context can ecological health be recovered. Only with this attitude can he maintain the self-control sufficient to the task of environmental control. He must see that he is responsible for utilizing his powers of control and communication to enrich the life of man and benefit the human community. To do this a new kind of human being is needed—a human being who, seeing God as Lord of matter and persons, can consider his cybernetic extensions into the environment instrumentally and can relate with freedom and concern to his fellowman.

Dr. Max Lerner spoke of this emergent man in an address at Monmouth College in 1965. This man would be

> . . . one who will not recoil from technology, who will not revolt against the machine, who will not become a machine-wrecker; one who will accept technology and try to work with it; one who will not recoil from power or from change nor from the reality principle in the real universe. One who will refuse to become dehumanized by the machine . . . one who will put out the antennae of sensitivity in order to pick up the new tremors in the life of the mind and the spirit . . . One who will know that man can become a monster if he forgets the nexus that ties him by human connection with his fellow beings. One who will recognize the great chain of being that ties all human beings together.[18]

The self-controlled man, the man who will not be dehumanized by or give superhuman power to his technology, is the only hope for the cybernated age. The imperative for cybernetic man becomes a willingness to submit to God's control so that his control might be humane.

A feedback network is the necessary prerequisite to individual and international control. Only as man, one being capable of transcending the fundamental dimension of control, submits to higher levels of control and communication will hope be possible. Here again, man must recognize his viability to cybernation and his transcendence. Sinsheimer, referring to the biological alternation of life, located both elements:

> . . . how can we possibly predict the ultimate consequence of our alternation of ourselves? Each small step will lead inexorably to another, in a cumulative, positive feedback mechanism to patterns of life and forms of knowledge and even systems of thought beyond our scope. We will have need of hope. . . . Man is psychologically the most plastic, least programmed animal; and by coincidence or by design he is self-aware. Thus he knows conflict, and thus he knows hope.[19]

Responsible control by man over himself and over social interaction must parallel his growing control over environment. Without this, man's invention shall overcome him. The source of this control is the transcendent God. Only in submission to this Greater can technological man achieve greatness in his control.

Goethe speaks poetically on this theme of control and subservience. In his poem *The Sorcerer's Apprentice (Der Zauberlehrling)* he tells of the young assistant to the master who tastes the sweet power of control one day and in the master's absence calls, with the proper incantation, on his broom to gather water. The broom pursues the task with great efficiency. Then the water tub begins to overflow, and the servant is nearly deluged, for he cannot recall the incantation used to stop the broom. The master returns at the point of disaster and summons the broom to its former submission.

An irresponsible servant cannot safely coerce his artifacts to his service. The message is a simple illustration of the dehumanizing potential that is present when control in man is not concomitant with cybernetic control over environment.

Cybernation also carries in its development the dehumanizing possibility of enslaving man to his extensions. Man can submit himself to this bondage as he locates power in and relinquishes trust to these cybernetic devices.

The possibility can be theologically stated in terms of idolatry or failure to see extensions as instrumentally indifferent in terms of power. Both of these dimensions are present in Paul's discussion of slavery in 1 Corinthians 6. Here again the context is dietary practice, but the implication is much broader: " 'All things are lawful for me,' but not all things are helpful [ουμφέρει]. 'All things are lawful for me,' but I will not be enslaved [ἑξουσιασθήσομαι] by anything" (1 Cor. 6:12).

The thrust of this passage is that power (ἑχουσία) must not be located in any subordinate, conditioned, or penultimate reality. When power is thus located in a preliminary reality, man submits himself to bondage. The artifacts of his making or the fabrications of his society can usurp this kind of power. Anything in the creation—possessions, objects of sacred trust, even "goods and kindred" *(Gut, Ehr, Kind und Weib:* Luther)[20] can become ways of life that enslave man: how much more the cybernetic extensions of his personality that control and manipulate his environment. These can so literally become a way of life that the intensity of relationship approaches the "one body" (ἓν σῶμα) identification that Paul analogically locates in the prostitute (1 Cor. 6:16).

The surrender of humanity at this point is urgently raised in the fields of cybernetic brain manipulation. Recent experimentation in chemical and electrical brain control highlight this. With chemical injections or feedback response from electrodes implanted in the brain, scientists have been able to control the behavior of rats[21] and even sidetrack a charging bull by radio control.[22] Neurophysiologists have shown that certain areas of the brain can be electronically stimulated to produce great pleasure.

The danger is particularly acute in this day because of implicit rejection of free will within the scientific disciplines. Every innovation in the field of human behavior control constitutes a further deterioration of respect for free will. Dr. Robert Sperry, the psychobiologist at California Institute of Technology, notes this threat: "Another serious threat to cherished images of human nature is the scientific rejection of free will. Every advance in the science of behavior, whether it comes from implanted electrodes, psychimetic drugs, the psychiatrist's couch, brain surgery, imprinting, or skinner boxes, seems only to reinforce the old suspicion that free will is merely an illusion."[23]

The cybernetic replacement of the mental functions of decision-making also has social possibilities of enslavement. The computers presently used in military decision-making are subject to this potential danger. Performing calculations that are highly complex with great speed greatly enlarges the possibility of error. In cases where the computer is programmed so that it implements its own decisions the problem is intensified. Perhaps the danger is not as dramatic as Eugene Burdick and Harvey Wheeler suggest in their political-science fiction,[24] but they rightly locate the dehumanizing, enslaving potential as decision-making is removed one or two levels from man's immediacy.

Henry Clark identifies another enslaving possibility

that computerization potentially holds. He sees a grading down of human value and creativity in the computerized society:

> ... the availability of the computer, or other technologically sophisticated devices, might lead to the adoption of faulty criteria of importance or excellence, thus seducing thinkers and scholars into investing their energies in trivial research or being satisfied with false solutions. The danger in any case, is that man might cease to be master of the computer and judge of its works, and become the slave of his invention, by whose insensitive standards he himself is judged.[25]

The danger of leveling down takes on degrading and dehumanizing potential as the computer and other cybernated devices are applied to education. Here programming standards have great danger of imposing order and correction to the detriment of creativity. Here again, control on the control becomes imperative to avoid enslavement. When technology invades private existence conformity is enforced, only marginal diversity is tolerated and society in general is reduced to the level of the least common denominator.

Another illustration of this is the feedback influence of ratings of television programs. Programming stresses the themes that return the greatest return in terms of viewing and supporting sponsors. When feedback results are rigorously followed (Nielsen ratings, etc.) only the lowest taste level becomes standardized.

A related problem that we face more and more is found in the computerized prediction that is used in the field of political polls and election returns. The public often feels as if it has lost control of its own decision-making privilege when the great N.B.C. or C.B.S. computers go to work in predicting election results. The first results come in from a small farm community in Maine. They are analyzed, sifted through the statistical information available on the voting patterns of that precinct, deviations from the expected pattern discerned, and the findings generalized so that the national results are stated. All this occurs only moments after the first polls are closed. The problem arises, of course, when people on the other side of the country, four hours behind in time, have yet to vote. It takes real political fervor to go out and vote for a man you know has already lost the election. Here we see how crucial it becomes to handle information responsibly so that the feedback loop is kept open and freedom in the decision-making process is not obliterated.

Hans-Eckehard Bahr has surveyed theologically the use of information in Christian history in his book *Verkündigung als Information.* He shows how in the openness of the democratic system it is vitally necessary to maintain responsible use of information. When the social structure demands an informed citizenry, and the channels of information are so highly sophisticated, checks and balances must be maintained to sustain the free flowing cycle of the feedback loop. If this freedom is not protected the worst kind of tyranny can envelop the most sophisticated nation. As Bahr has noted: *"Nicht die Freiheit der Meinungsäßerung also, sondern die Freiheit der Meinungsbildung ist heute gefährdet."* (Both opinion changing and opinion making are threatened.)[26]

The downgrading of informed responsibility is one of the great challenges of the cybernated era. The mood is reflected in the lack of interest on the part of a demonstrating student generation to demonstrate for the right to vote at age eighteen. The responsibility is abdicated; they do not crusade for the vote, for one reason among others, simply because this generation feels it is futile to exert influence on a political system whose dynamics of change lie at another level of the communication network. The restriction that irresponsible use of information can bring is only beginning to be discovered today.

Herbert Marcuse, in a very important study, labels this whole intellectual and cultural phenomenon a reduction to "one dimensional man." Urging the intellectual to side with the worker in a way that is influencing students across the Atlantic more than it is in his own country, he seeks to discern the rationality of the technological society in order to restore the creativity that he finds obliterated. Here, because of control through technology, the entire cultural expression is dominated and downgraded to a point of mediocrity. ". . . the way in which a society organizes the life of its members . . . tends to determine the development of the society as a whole. As a technological universe, advanced industrial society is a political universe, the latest stage in the realization of a specific historical project—namely, the experience, transformation, and organization of nature as the mere stuff of domination."[27] An interesting cultural observation is that Eastern thought has always reacted to this kind of slavery to one's artifacts. For this reason with some exceptions the Eastern world has had an anti-technological spirit. The Taoists, says John C. H. Wu, "were convinced that all the troubles of the world had their origin in the love of knowledge and the employment of artifacts."[28] Laotse expressed the spirit simply when he advised:

> Let there be a small country with a small population. Even though there should be mechanical contrivances requiring ten times, a hundred times less labor, there would be no occasion for using them. Cause the people to love life so that they would

> hesitate to run the risk of death by moving to distant places. Boats and carriages there may still be, but there will be no occasions for riding them.[29]

This idyllic longing for the simple life, though understandable today as when first written, is naïve in terms of the responsibility that the contemporary world requires. Wherever this kind of escapism from the world is present there is usually a gross deprivation and social injustice while, in fact, only the privileged enjoy the carefree life. The simple agrarian existence has never been the paradise of the cotton-picker.

Our concern takes on ethical significance at this point. We cannot retreat to the simple pastoral life. Technology and cybernation are gifts given by God to the world through the ingenuity and industry of man. You cannot go back to some simpler past. What can be done, must be done; done responsibly. Technological gifts are not to be rejected; rather they are to be seen in the context of stewardship. These developments are to be welcomed, not worshiped; utilized, but not served. The cybernetic extensions of man are servants not masters. Herein lies the ethical challenge. Norbert Wiener has summarized in this way:

> The problem, and it is a moral problem, with which we are here faced is very close to one of the great problems of slavery. We wish a slave to be intelligent, to be able to assist us in the carrying out of our tasks. However, we also wish him to be subservient . . . How often in ancient times the clever Greek philosopher slave of a less intelligent Roman slaveholder must have dominated the actions of his master rather than obeyed his wishes! Similarly, if the machines become more and more efficient and op-

erate at a higher and higher psychological level, the catastrophe foreseen by Butler of the dominance of the machine comes nearer and nearer.[30]

In the intellectual history of Western civilization one of the perennial challenges of the philosophic efforts to objectify reality, of which cybernation theory can be seen as an ultimate attempt, is that man, the subject himself, becomes objectified in the process. Ethically speaking the integrity of man, both in his intellection and his action depends on his ability to resist this objectification and maintain the dialectic of relationship. One of the weaknesses of the minimal theoretical literature that exists on the theme of cybernation is the refusal to recognize this crucial distinction. Until this theoretical problem is recognized it is hardly likely that there can be significant criticism of practical cybernetic implementation.

The possibility of slavery to the machine is one aspect of the larger threat which is the impersonalization of man in society. To remove man, understood as subject, both from the personal level of understanding and the understanding that is current in society, replacing this with a mode of understanding which sees him as object, is one of the lingering ethical problems of cybernation.

If slavery can dehumanize man in terms of control, impersonalization accomplishes the same thing by devaluing man's personality. The indictment made at this point comes from many quarters. Most often it is addressed to the industrial-technological movement of history in which we stand. Although cybernation develops within this historical milieu, it raises the problem in a unique way. While the threat of depersonalization is present even in the primitive society, in the cybernated society life's vocational meaning tends to be defined impersonally. Not only is this the private mode of understanding but it is buttressed and generalized in the society. Not only does society tend to dehumanize man by its patterns of life and work; man himself begins to conceive of his existence functionally and impersonally. He begins to take pride in his functionalism, his typification. Karl Jaspers analyzes this existential dimension of depersonalization:

> The new, non-personal man moulded by the concept of technological existence, comes into being. He sees himself as a type, not as an individual, unsurpassable in the skill and sureness of his function; he prides himself on obedience, because he lacks any desire for an existence of his own; he may secretly esteem himself as a great and superior to all that is personal, because to be replaceable at will by others intensifies his own self-consciousness. He hates solitude, lives with open doors, and craves no privacy; he is always available, always active, and true to types; lost indeed if left to himself, but indestructible, because his replacements continually grow up after him. Personality becomes an old fashioned, ridiculous concept. Being a type gives strength, satisfaction, and the consciousness of perfection.[31]

This type of man, as typical in the cybernated era as in the Nazi era of Jaspers' recollection, is basically unethical because he does not stand under the terror of the dialectic which keeps his life in the perpetual tension that as a son of God he has in the resistance to objectification. He sees himself as perfect because he meshes so well into the productive process. His efforts are perfectly synchronized to the energy feedback loops in which his life is situated. He is "well rounded," like a ball bearing.

This new man is dehumanized because he has surrendered his freedom of decision and responsibility. He has escaped his freedom by relinquishing his destiny to functional definition. His life style is characterized by the words "he fits well into the system." The cybernetic era carries this dehumanizing potential forward because of the accelerating power of its organization over life. Man does not withhold his surrender lest the goddess of success pass him by.

Erich Fromm labels this man *homo mechanicus.* He is the man who has surrendered his life-orientation for the death-orientation of things. "By this [*homo mechanicus*] I mean a gadget man, deeply attracted by all that is mechanical, and inclined against that which is alive."[32] Who can fail to see the fascination of modern man with his ambulating extension, the automobile? Studies in Germany and America have even revealed a disintegration of the sex impulse and other basic human drives under man's compulsive attraction to his cybernetic extensions. One could also note his obsession with television, his visual extension, important enough to merit twenty-five hours of his time each week on the average. Fromm links this to destructive motivation: "Homo mechanicus becomes more and more interested in the manipulation of machines rather than in participation in the response to life. Hence he becomes indifferent to life, fascinated by the mechanical, and eventually attracted by death and total destruction."[33]

Individually man can be dehumanized in the cybernation process. The terms we use to describe the deindividualized man in our society are mechanical terms. "He is a well-rounded man." Just like the ball bearing he rolls smoothly with the machine. Or we say "he is adjusted," another technological term. "Man," says Tillich, "resists objectification; and if his resistance to it is broken, man himself is broken."[34] Depersonalized man ceases to be a man and poses a threat to the fabric of the social order.

The dehumanizing potential of cybernation takes on social dimension at this point. Not only does depersonalized man threaten his society, but a society that has been depersonalized allows him to do it and enhances these negative impulses.

The society that has been depersonalized cannot help turning out depersonalized individuals. Mass society, which is the natural outcome of the cybernetic emphasis on organization, inevitably leads to a depersonal emphasis in relationships. Thielicke points up that this attitude pervades even man's recreation. Referring to the fear that man's "humanness" is being threatened by technology he writes: "I am also thinking of the tendency to deal with people in the mass that comes with technology, and further of the degeneration of recreation which is a concomitant of this dealing with people in masses and is becoming less a means of gathering oneself together than a diversion and a distraction."[35]

Much has been written of the depersonalization of society. The futuristic vision of Orwell's *1984* or Huxley's *Brave New World* has assumed tangible shape in the cybernated age. Kierkegaard reacted against certain features in mass society in *The Present Age.* The theme has developed through the social protest literature of the last century. Kafka revolts against depersonalization and anonymity in mass society by refusing to name two characters in his later novels. The persistent underlying theme is that society is distorted as individuals are dehumanized. The perpetual danger of *Gesellschaft* (the organized society) is the disintegration of not only *Gemeinschaft* (organic community) but also individual personality.

The cybernetic revolution accentuates this danger. The character that society takes is not only organized, it is self-organized. Men are not only free, they are estranged from the productive process. The highly specialized division of labor has reinforced this depersonalization. As impersonalism occurs, power and decision-making tend to be located in the hands of smaller and less representative segments of the society. Although participatory democracy is made fully possible with electronic media, highly organized society often takes on conditions of tyranny that parallel primitive society. When the flow of information is so massive and the press of the decision-making demand is so intense the interrelatedness of all kinds of decisions necessitates the ability of a few to stand at this point of convergence and make responsible decisions that are reflective of the desires of the populace. The best illustration of this concentration of responsibility occurs in military situations where the few strategists hold responsibility not only for the subordinates but for the nations represented in the conflict.

Père Dubarle, a Dominican friar, has written of this danger in *Le Monde.* In reviewing Wiener's thought he notes the danger of mass exploitation by the few who have the privilege of decision-making. The essence of the unusually urgent article is that in the cybernetic age man must either willingly subordinate himself to the higher decision-makers through conscious relinquishment of these prerogatives or go the way of passive submission in the manner of sheep. In order to avoid the tyranny of the great world state that subordinates the inhabitants to some perverse will, Dubarle suggests, "Perhaps it would not be a bad idea for the teams at present creating cybernetics to add to their *cadre* of technicians . . . some serious anthropologists, and perhaps a philosopher who has some curiosity as to world matters."[36] The depersonalizing possibility has individual and social dimension. It therefore becomes imperative to focus ethical concern at both levels. A pure individualism will not speak to the deindividualizing social structures nor will a socio-cultural ethic meet the existential challenge.

A final dehumanizing potential must be located in the temptation within the cybernated era to secularize God or to divinize the process. We noted initially the idolatrous situation which, because of either mis-emphasis, creates the conditions for dehumanization.

Ronald Gregor Smith has shown that the transcendent is by definition that which man cannot completely possess.[37] When man assumes that he can do this or when he psychologically convinces himself that he has achieved this possession he has dehumanized himself. Without transcendence there is no meaning to temporality. Many insist that transcendence must be defined secularly today. The motivation for this desire is partially contingent on the glorification of technological progress. When man can control and utilize his environment in the sophisticated way of which cybernated man is capable he tends to deny the validity of transcendent reality. The system is self-contained, self-explanatory, and therefore autonomous.

Speaking from the biblical-theological context this is invalid. Thielicke points up the way in which Psalm 104 acknowledges the I-Thou structure of nature: "Psalm 104 . . . describes the processes of nature not as a self-contained, autonomously functioning mechanism, but rather as a drama in which God is acting at every single moment, and without his sustaining preservation and intervention the world would immediately collapse and, to speak in modern terms, the laws of nature would disappear."[38] Here within the context of faith is found the proper view of nature that alone can sustain humanity within the cybernated process. If the process is given ultimacy or if

God is defined secularly the dehumanization possibility arises. Our task therefore is to recognize God in the secular process, even in the cybernetic phase which is our history. But we must ascribe to God that transcendence that alone recognizes his sovereignty and alone protects man's humanity.

Dr. John Mackay, an eminent scholar-churchman, has recently written appropriate words of summary. Speaking critically of those who defined God secularly he says: "The reality of God's presence or its equivalent should, therefore, be sought in the achievements of that august new divinity, technocracy. Now, let it be quite clear, God is present in the secular city. But he is there as sovereign Lord and concerned lover, not as prisoner or anonymous tenant, and still less as a cybernetic potentate."[39]

The first three parts of this book have developed the context for a definition of responsibility in an age of cybernetics. We have shown the ambiguity that characterizes the context of our problems by developing the humanizing and dehumanizing potential. Now our task is to speak of responsibility in the midst of this cultural ambiguity. In this chapter we have provided a working definition of cybernetic dehumanization by relating the concept of idolatry to our problem. We have noted the dehumanizing potential in the lack of concomitancy between self-control and environmental control; in slavery to one's artifacts; in depersonalized man and society; and finally in a secularized restriction of God. With this preliminary work done we can now pursue in depth the concept of man as co-worker with God in subduing and controlling the earth and see how responsibility is derived within this understanding.

IV

Cybernation as Co-creation

While the world stood as witness, engaged through electronic circuitry, on the 20th day of July, 1969, man took his first step onto another planet. What measure of act was it? Man had channeled propulsion and communication energy in pioneering ways. Was this exploit an act of Promethean defiance—a grasping of the fires of the gods—or was it an act of sharing of the divine activity? Is man's subduing of the cosmos violation or co-creativity? This section responds to these themes.

We have defined and delineated the problem. We have noted the ambiguous ethical situation that characterizes cybernation in its cultural setting. We are now ready to explore the problem of responsibility. The last three chapters will undertake this task. We will note the way that ethical responsibility is derived, directed, and actuated within cybernation. This chapter will show how ethical norms are derived within the basic man-world relationship as this is expressed in the biblical concept of man as the co-worker with God in the task of subduing the earth (Gen. 1:28).

Initially we must admit that it is not proper to seek within the Scripture moral directives stated specifically for our technological age. This attempt would prove futile for two reasons. The first concerns the meta-scientific character of the biblical documents. They are not primarily concerned with an ethical-scientific analysis of a particular time. Their emphasis is rather historical and relational. Secondly, we must acknowledge the radically new time in which we live. The understandings of man and community, world and cosmos, have been fundamentally altered in our time. Where the cosmos was formerly conceived in static, Parmenidean ways, we now perceive reality in historical, developmental ways.[1]

The fruitful approach to our problem, therefore, is to locate the man-world relationship in the Scripture. Then, having noted the way that responsibility is derived within that relation, we can apply this to our task, even though the network of relationship is much more complex within cybernation. The chapter will be developed in four steps. First, the man-world relationship will be surveyed historically as this provides the context for responsibility; then the biblical theme of co-worker *(Gottes Mitarbeiter)* in creation will be developed. Responsibility will be defined and described against this background. Finally, this will be related to the cybernated era.

The evaluation of responsibility in the age of a cybernetic man-world environmental situation must remember and rediscover the basic man-world situation of Christian history, beginning with biblical thought. Only therein can the Christian derive the ethical norm. Our initial task, therefore, is to trace the interpretation of this basic man-world relation specifically at the point where it shapes responsibility, first in the Old Testament, then in the New Testament and early church, in the Reformers, and in modern theology.

Surprisingly enough, the subject of creation and

man's responsible place within it is a minor theme in the Old Testament. Von Rad notes that the only treatment is found in the later texts, mainly Deutero-Isaiah, the priestly document (P), and a few Psalms.[2] The reason for this is probably the centrality of the saving-acts of God consciousness in the Hebrew experience. The import of this point is that in the Hebrew documents creation cannot be understood apart from the saving history *(Heilsgeschichte).*

Three motifs in the Old Testament creation narratives have direct bearing on the derivation of man's responsibility with reference to his subjugation of the world. We consider the dominion motif, the co-worker motif, and the redemption motif.

The central note of the creation narratives is that God brings the cosmos into being through his Word (Gen. 1; John 1:1-3). The world is not the result of creative conflict of eternal tension, whether it be of beings or power; it is sheer creative act ex nihilo. Therefore, the divine is not intricated in the creation as in demiurge or emanation theory. God is radically sovereign over the world; he possesses it; he has full dominion.[3]

The import of this view is that only God has enduring reality. The cosmological significance of this, as Von Weizsäcker has noted, is that the world has been deprived of its divinity since it is a creature contingent to, rather than essential to, the Creator.[4] The dominion of God necessitates the disenchantment of creation. The world is subject to the creative coercion and manipulation of God. It has no fundamental autonomy in terms of self-sustaining and self-ordering power. The point was precisely made by Newton who, though relegating God transcendent to natural process, located in God the power that sustains natural law, continually corrects irregularities in cosmic process, even preventing universal collapse under gravitational pressure.[5] With Einstein's thought, the scientific accuracy of Newton's emphasis is, of course, fundamentally altered; but the glimpse of God-world relationship is pointed. Normal process, though primarily inexorable by God's intention, is pliable to creative act in sustaining, correcting, and developing ways. God's dominion over creation implies his will to build the earth—to develop and to perfect it. The radical detachment of God and implicit isolated autonomy of the creation in Gnosticism and Deism, for example, is the very antithesis of the God-world relationship in the Old Testament. The implication of this point for cybernation is that man, related responsibly to God as a co-worker in subduing the earth, becomes companion with God in creative activity.

In order to derive responsibility within cybernation, we must now relate the dominion motif in the Old Testament to man as the co-worker. This will be the second plane of relatedness as we seek to derive man's responsibility. We have noted the God-world relation. We now analyze how God is related to man; from this we will move to the man-world relatedness.

Von Rad makes the interesting observation that in the Old Testament the understanding of the Jahwist tradition was not to anthropomorphize God but to theomorphize man.[6] The importance of "glory of God" motif in Ezekiel 1:26 and even the *imago dei* motif in Genesis 1:26 is that man is fashioned in distinction from the created world, outside the sphere of the created, in the sphere of God. Not only does man in creation have relatedness to God, but he shares intrinsically in the attributional creativity of God. Freedom, will, transcendence, purpose, rather than being human attributes projected on God, are divine capacities that man shares in his unique place in the creation.

Man is placed on the earth as God's co-worker. Man becomes the expression of God's dominion over the earth. This commission to subdue the earth (כבש רדה) is the extension of God's Lordship over creation. Man is the partner with God in manipulating and constraining the earth to fulfillment. This function is related essentially to God's dominion and man's image. Von Rad summarizes: "What is crucial about man's image is his function in the non-human world."[7]

Here we see the relationship within which responsibility is shaped in the Old Testament. At the juncture where man shares the creative work of God, he works as the expression of God's dominion. His work is not free creativity. It is derived from the purposes of God. To define this responsibility, we must note one other motif of Old Testament thought: the redemption thrust of God's will in creation. We leave the co-worker motif at this point; it will be more thoroughly developed later in this book.

Responsibility can be seen most clearly in its derivative character in the Old Testament if we note the soteriological or redemptive motif in creation. Here we find the (τέλος), the purpose of creation, and thus the character of man's responsibility as God's co-worker.

According to the Old Testament, the cosmos is created with a purposive destiny. It is not a happening or an accident; it is imbued with creative origin and direction. In the Genesis narratives the world is disclosed as the setting of man's life and the stage of saving history *(Heilsgeschichte).* In Deutero-Isaiah, a new element is unfolded. Here in Isaiah 44:24, for example, the creation itself is seen in terms of a saving event. The term for Yahweh is "the Redeemer and the Creator" (Isa. 45:12; Jer. 27:4-5; Ps. 77:16-17).

Von Rad suggests that a soteriological understanding

of creation is basic to both J and P.[8] The creative act of forming the world and the choosing of Israel are coincidental, integrally related events.

It is crucially significant that the link of Creator to creation is through his Word. The radical meaning of ברא is that the Word is the only continuity between God and world. The world therefore originates in, exists with, and proceeds to the purposive expression or will of God.

The import of this soteriological conception of creation is that the world is going someplace. It is developmentally ordered. It is in process. It is not already perfect, but, like man, it exists in a state of striving for fulfillment. That fulfillment is none other than the redemptive consummation of God. All creation struggles to achieve and perceive the salvation, the wholeness, the intention of God.

Man, as co-worker with God, participates in this redemptive venture. Man's renewal in the purposes of God is accompanied by the renewal of the creation as man subjugates the earth through control and communication. We will note this theologically when we discuss the redemptive co-worker motif in the New Testament. Cybernetically, we will relate this point at the end of this chapter.

In summary, we note that the Old Testament speaks of the man-world relation in the context of a God-world, God-man relation oriented in the dominion of God over creation, structured cooperatively and motivated redemptively.

In the New Testament community the man-world relation is also a minor theme. Against the sweeping, cosmic drama of salvation with the pivotal emphasis on the cross of Christ, followed by the eschatological urgency of the early church with its emphasis on the passing character of this world, the man-world relation pales to near insignificance. However, the redemptive motif in the New Testament writings bears directly on our theme. To summarize this motif in the New Testament Scripture and make reference to the creation theology of Justin Martyr, Irenaeus, Tertullian, and Origen is the purpose of this section. The attempt is to abstract from these an early Christian conception of the man-world relation.

The redemptive motif appears most clearly in the writings of Paul. The book of Romans reflects on the fact that the creation itself is subject to the same futility that man is. The world groans and travails as it longs for fulfillment. The future glory of the world is related with the renewal of man in the whole cosmic drama of redemption. Paul speaks in Romans 8 of the eager longing of the whole creation—all things animate and inanimate—for that day of release from corruption and decay.

In a very contemporary mood, Paul speaks of the pathos of a cosmos that has an internal perception of its destiny, not yet revealed. The present agonizing is, for Paul, eschatological anticipation of redemption. George Buttrick summarizes: "Here [in Rom. 8:18 ff.] Paul alludes more particularly to the sorrow of nature. He thinks of the suffering of animals. The weak devoured by the strong—of the ruthless destruction of plant life, of natural catastrophies of all kinds; he listens, it is not fanciful to suggest, to the crying of the wind and the sea . . ."[9]

In crying for release from this pain, this incompleteness, the created order itself participates in redemption. The relation of this motif to the dominion mandate over the earth in Genesis, to the Fall, and to the anticipated redemption, relates this theme to cybernation. Control and communication of environment not only accentuates this cosmic trauma but intensifies eschatological expectation.

We must carefully differentiate at this point Christian redemptive eschatology from uncritical evolutionary optimism. In an important new book, *The Biology of Ultimate Concern,* geneticist Theodosius Dobzhansky calls on this passage from Romans to justify human progressivism. Dobzhansky's comment on Romans 8:19-22 is: ". . . since the world is evolving it may in time become different from what it is. And, if so, man may help channel the changes in a direction which he deems desirable and good."[10] Certainly the traditional thought of inevitable progress or determinism needs to be balanced by consideration of the freedom of man in the world. But it must be affirmed that redemption, in the Pauline sense, of both creation and man, is of divine initiative and impulse, with man as co-worker, rather than the reverse.

The period of the apologists represents the first dialogue of Christian faith with the prevailing intellectual milieu. The writing of Justin Martyr typifies this period when Christian theology seriously engaged the prevailing cultural world view. Eusebius suggests that Justin lived at Rome "in the dress of a philosopher."

Justin's theology of creation and man's place in it is colored by his Platonic bias. The natural world is a low order emanation subject to the eternal process of change and decay. Justin departs from this emphasis by asserting that the world, created by God who is alone (unregenerated) begotten, is (generate) begotten, having existence which is derived, not independent. The thrust of the point that the world is made by God is not only that it is radically other, not to be confused with God, but that the world has origin and destiny through God and is ordered by creative fulfillment. God has brought order from chaos through the logos; and the logos gives to the world a continuous "becoming" character. God is not subject to this developmental process, but the world is.

The legacy of the work of Justin Martyr and the apologists has maintained for Christian theology the redemptive character of creation. The world is not a static entity, but a process of becoming. Cybernetically speaking, the world is transformable.

The French Dominican, J. D. Chenu, contemporizes this conception in an article, "The Need for a Theology of the World": "God did not create a fully developed universe and then place man over it like an angelic spirit over heterogeneous matter or like an alien spectator before an alluring and overwhelming landscape. God has called man to be his co-worker in the progressive organization of the universe."[11] Here we see the initial significance of the *imago dei* character of man. Man becomes the extension of God in the world to constrain the world toward its destiny. Chenu concludes that "man is precisely and primarily the 'image of God' in this association with his creator by which he is the master and constructor of nature."[12]

Serious questions arise when we posit man as co-creator of the universe as Chenu proceeds to do. The logic of reasoning from this basic developmental conception of nature to man as co-worker is legitimate, even necessary, but co-creation raises questions out of the scope of this paper.

In this discussion of Justin Martyr we have noted the relevance of the redemptive, developmental concept of nature for a man-world relation which locates man as co-worker in the organization or ordering of this development. Cybernation identifies man as the locus of organization in terms of feedback and control.

Irenaeus, Bishop of Lyons, though the follower of Justin Martyr, brought radical correction to his thought as he articulated Christian orthodoxy against the challenge of Gnosticism.

The Gnostic world view shared certain similarities with the neo-Platonic. There was the original, immortal world, the πληρώμα (the fullness) where perfection and incorruption reigned. Then there was the material world, where corruption and evil were intrinsic reality.

Irenaeus follows Justin Martyr's distinction between Creator and creature and the concomitant generate and ingenerate existence. Yet Irenaeus, motivated to defend God's sovereignty in creation and protect against the Gnostic dualism isolating God from the created world, affirms God as sole and direct Author, Sustainer, and Redeemer of the cosmos.

Irenaeus repudiates the idea that the created world is the result of God's molding and manipulation of resistant, evil matter. God called the world into being. There is no material cause extraneous to the divine will. "Men indeed," says Irenaeus, "are not able to make anything out of nothing, but only out of material which lies to hand. But God is greater than men in this first regard, that he himself summoned into existence the material of his creation when before it had not been."[13]

All levels of creation are subject to the redemptive action of God, redemption in Irenaeus not meaning extrication of spirit from matter but rather transformation of matter itself. God is sovereign. He can transform corruption into incorruption; he can resurrect the body.

The significance of Irenaeus is seen in the world view which became orthodox partly as a result of his apology. The world is good, not extraneous to God and devoid of his sustenance. The world is in the process of becoming according to his will. Man stands in creation as part of the redemptive process which is transforming all creation. Man is the apex of created reality, as well as participant in the first fruits of redemption in Christ. He stands then as transformed and transformer while God's new reality organizes all creation.

Tertullian must be mentioned here because of the divergent view of man's position in the world which he offers. Tertullian is the first Christian theologian to write in Latin. His world view, though Greek-flavored, is Roman. He was Roman particularly in the rigorous distinction in his thought. Because of a presupposition of revealed truth, he made radical distinctions: things were either fully right or wholly wrong according to revealed doctrine in the Jewish Scripture, along with the evangelical and apostolic writings. To this authoritarian base was added a radical mistrust of philosophy. The wisdom of man is fallen and philosophers are the "patriarchs of heresy." "What has Athens to do with Jerusalem, the Academy with the Church, the heretics with the Christians . . . of what importance is it then, that there are men who have set forth a 'stoic' and a 'platonic' and a 'dialectical' Christianity? We for our part have no need of curiosity now that we have found Jesus Christ, and no need of searching after we have found the Gospel."[14]

Tertullian unfolds his creation theology in the treatise *Against Hermogenes.* He affirms creation *ex nihilo* is radically centered in the creative logos. "What we worship is the only God, who by his Word . . . drew out of nothing, for the glorification of his majesty, the whole immense system."[15] He becomes the first in a long train of Christian theologians to affirm the radical transcendence of God. The "otherness" and "over-againstness" of God to the world slights creation and incarnation theology. The paradox of this polarization is similar to what has occurred in contemporary theology; radical transcendence collapses into secularism. Just as sacred/secular polarity leads ultimately to a "death of God" mentality, so with Tertullian "otherness" almost collapses into materialism. "Since . . . the universe belongs to the Creator, I see no room for any other God. All things are full of their Author, and occupied by him. If in created beings there be any portion of space anywhere void of Deity, obviously what it is empty of is a false God."[16] Tertullian attempts to find a God who is engaged in the events of history and related to the development of the world. His presuppositions, however, forbid such a position. Tertullian's value for our discussion is that the radical view of God's sovereignty demands a high view of creation and man's place in it. When sovereignty implies deprecation of creation the doctrine itself collapses.

Tertullian first articulates a line of creation theory which has remained forceful in Christian theology. Such theory brings correction to any naïve progressivist creation theory to which this thesis is particularly vulnerable.

Brief mention must be made of Origen because of the significance in his world view of the synthesis of Christian theology and Platonic metaphysic. The entire *Weltanschauung* rests on a doctrine of logos. Man, created by God's Word, is informed at the highest level with rationality which is participation in logos. The world, too, is created by the Word of God so that it is integrally related in a pattern of created intelligences.

Christ, the incarnate logos, becomes the pivot point where God, man, and world intersect. As man participates in the Son's unity with the Father, he becomes part of the goal toward which the whole creation is moving. Here, again, is the dynamic view of created reality that is lacking in Tertullian. The logos becomes the active, informing principle in the movement of all creation toward its destiny.

This ordering principle, given personal redemptive significance in Christ, is highly relevant to our thesis. As order is brought by the Word of God from chaos and disorder, his will is accomplished. As communication and control constrain the created world to God's purposes, his work continues. The cybernation phenomenon extends the reign of rationality into the cosmos. It orders reality under unifying principles of reason. The crucial question, therefore, becomes the means and ends, the content and consummation of man's newfound abilities to control the world. If his motives are brought to coincide with divine will, he surely stands within his creative activity.

The early Christian theologians, both Latin and Greek, contended for a dynamic creation, postulated on the sovereignty of God, with man in a pivotal position in the creative process. The point was made variously, either in a theory which resulted from synthesis with philosophy or in reaction to prevailing secular thought. Whether the argument was creation as over against Creator, or logos principle, the same dynamic view of creation with man pivotal to the process was maintained.

We must now refer to the thought of Calvin and Luther as they speak of man's responsibility in the ongoing process of the world. Here again we must acknowledge that this is a minor theme in both Reformers' thought. Calvin's theology accentuates the glory of God with the world seen as instrumental to incite man's praise of the sovereign God. In Luther, a soteriological emphasis, with the dialectic between grace and faith, along with an Augustinian polarization of sacred and secular, places man's responsibility in the world in a more remote position.

Recent scholarship, however, has recovered the strong emphases on creation and man's part in the process of subduing the earth in both Calvin and Luther. We abstract the thought of Calvin, then Luther at this point.

For Calvin, "the whole order of this world is arranged and established for the purpose of conducing to the convenience and happiness of man."[17] Man stands at the pinnacle of this creation which exists solely for his life. Man has "a singular honor, one which cannot be sufficiently estimated, that mortal man as the representative of God has dominion over the world, as if it pertained to him by right."[18] In his commentary on Genesis he relates man's dominion to the *imago dei* and the dominion of God. "It is thus too that man exercises rightly his dominion over the earth—if he gladly and gratefully submits to the gracious dominion of God over all, for then his dominion over the world becomes part of the way in which he as man images the glory of God."[19]

The redemptive motif controls Calvin's view of nature. Located in Christ, redemption is the informing principle of cosmic development. "Christ . . . the lawful heir of heaven and earth . . . has not as yet actually entered upon the full possession of his empire and dominion."[20] In Christ all creation is striving for and shall achieve consummation. Calvin relates the accomplishment of dominion to the apostolic "every knee shall bow."[21] For Calvin, dominion over the earth refers to the subjugation and control of animate life for man's benefit. Dimensions of dominion possible today, the harnessing of energy, the electronic joining of person and event, the shrinking of space and time, were unimagined. Yet his theology has continuing significance. As cybernetic man participates in the teleological dynamics of creation, i.e., expression of glory to God, he participates in the most powerful energy dynamics of the universe. If his scientific mastery of nature is motivated by moods characterized by Calvin as awe and humility, rather than a usurpation of the divine prerogatives, that science will be richly rewarded and that dominion will participate in the divine dominion.

Luther, like Calvin, perceived the creation to be the panorama where God portrayed himself for man's excitement and the elicitation of man's praise. In a note in a volume of Pliny written in the last year of Luther's life, he said: "All creation is the most beautiful book or Bible; in it God has described and portrayed himself."[22] From Luther's earliest years all the resonances of nature signaled the terror and glory of God: the storm, the clouds, the lightning, and the birds. The morning dew and the night shadows both reflect God's presence.[23] Luther stood as a primitive man with fear and primordial awe before all creation.

In terms of transcendence and immanence, Luther cannot be conceptualized according to our present catagories of thought. With Nicholas of Cusa he pondered the creation and found God infinite in externality to the world; yet his internality pervaded and undergirded the most fundamental reality. God "is a supernatural, inscrutable being able to be present entirely in every small kernel of grain and at the same time in all, above all, and outside all creatures."[24]

Because of Luther's emphasis on radical justification and his postulation of the total depravity of the secular, the literature does not deal at great length with man's "being in the world." In *Luthers Lehre von den Zwei Reichen in Zusammenhang Seiner Theologie* Günther Bornkamm also accents the difficulty of finding passages dealing with man's relationship to the natural world. He has pointed out, however, the importance of the complexity of the two kingdoms concept, showing that it is not a simplistic extension of the Augustinian polarization of sacred and secular.[25] In the tension between the temporal and spiritual; between the *"Reich der Welt/Reich Christi,"* we find Luther affirming the importance of the secular world in redemption. For Luther the secular is posited with high order value that is emasculated in the cosmology of Lutheran orthodoxy. The essence of the two kingdoms tension is between law and gospel rather than the popularly caricatured sacred-secular tension. Luther, like Calvin, affirms the participation of the secular world at the heart of the cosmic redemptive process active in Jesus Christ.

Accenting this affirmation in Luther, enhancing rather than bankrupting the significance of the secular, is his thought of the co-worker. In the good study by Von Martin Seils, this theme of man's mutual work with God in Luther is pursued.[26] Man is *"Mitarbeiter"* with God in the salvation process. Seils develops the theme entirely in the soteriological context. The openness of Luther's thought at this point, however, implies the viability of the system to man-God cooperation in world subjugation, since this world, too, is the arena of saving activity.

According to Seils, the primacy of God's initiation and sustaining of cosmic process is maintained in Luther. Even when *"cooperatio"* is referred to *"mit dem Menschen,"* the inference *"durch den Menschen"* is implicit. At the same time, the necessity and integrity of man to the process is maintained by the *"mit"* implicit in the *"durch."*[27]

In Luther the identification of God's work with man's is prevented by the basic world view. For Luther, a fundamental polarity is maintained between God's government *(das Geistregiment)* and man's government *(das Welt Regiment).* Based on the New Testament distinction between gospel and law, flesh and spirit, the polarity is thoroughgoing; no confusion in the area of secularization is possible. In social theory the doctrine of the two kingdoms extends this bifurcation of reality. The theology of the cross accentuates this polarized tension. "The work and creation of Christ does not appear to be anything outwardly *(foris),* but its whole structure is within, before God, and invisible."[28]

This polarization, however, heightens Luther's estimate of the secular. Christian existence is located in the world *(in orbe terrarum).* The Christian man participates with God in and is subject under God to the *"weltlichem Regiment."* Man's calling *(der Beruf)* and estate *(der Stand)* in the world are orders of responsible existence. These secular orders are not extraneous and peripheral but basic as in responsible existence man participates in God's plan of cosmic redemption. If we can extend Luther's understanding of world government beyond the socio-political sphere to the cybernetic ordering of the world, we see the relevance of his thought. Luther links man's reason, his rational structures of decision and planning, to the process of man-God cooperation. Man participates in the jurisdiction of God's *"Welt Regiment"* through his reason. Commenting on Genesis 1:28 in a sermon on Matthew 6, Luther illustrates this position with regard to marriage: "The estate of marriage *(der Ehestand)* is a worldly, outward thing and . . . belongs to magistracy to rule *(zur Oberheit Regiment).* As such it is entirely subject to reason. Accordingly, one should allow to stand whatever the magistracy and men of wisdom conclude and order about it on the basis of right and reason."[29]

In conclusion, Luther emphasizes man's place in ordering reality. Man is responsibly free to structure and plan his social and technical life. In this undertaking he is engaging in *"cooperatio"* in the redemptive sense. Though secular reality is always conditional, man, because of his rational endowment, is participant in the divine task of world ordering. God, the Unconditioned, is the Impulse and Designer of world order. He can work alone *(allein Wirksamkeit),* but he chooses man as partner and the process is one of mutuality *(Zusammenwirken).*[30]

A conclusion of this synoptic overview of the man-world relationship as it appears in the history of theology must account for the radical change that occurs in the modern period. A fundamental shift occurs after the Renaissance. The significance of man's place in the world is heightened by the humanism of the Reformation and post-Reformation period. In the nineteenth century the modern ethos is fully explicit as man's place in the cosmos is glorified to the point of distortion. In Marx and Darwin the emphasis is made on man as the director and creator of world development. In the theological literature Feuerbach and Schleiermacher, in radically different ways, locate disordinate cosmic power in man's ability.

In *The Essence of Christianity,* Feuerbach notes with classic skepticism how "the doctrine of creation . . . arises only on that standpoint where man in practice makes Nature merely the servant of his will and needs, and hence in thought also degrades it to a mere machine."[31] Feuerbach, though puncturing the anthropocentrism of that century's theology, is only driven to a more pernicious subjectivism himself. Karl Barth called on Feuerbach's critique of naïve nineteenth-century optimism in his *Die Theologie und die Kirche* in 1928. Feuerbach was reacting to the pretense of Hegelian thought which reduced God to the hypostatization of the self. God thus became the personification of man, who was the measure of all things. In Feuerbach this reduction is radicalized so that man in his creativity and power becomes the center of all religion.[32] The Jewish-Christian conception of creation is just another wish-projection, an illusion.

Schleiermacher's basic view of man in the world also fails to maintain the fundamental distinction between God and world. In the *Dogmatik* (§§ 46-47) he elucidates his views on creation which proceed from his radical immanentism. In God, distinction cannot be made between potentiality and actuality. Schelling noted that pantheism is present whenever the relation of God to world is not distinguished in terms of possibility and reality. Schleiermacher loses precision at this point. The doctrine *Creatio Ex Nihilo* is valueless for him. The relevant implication of his thought in this essay is that nature and God collapse into confusion. Natural law and divine providence are identified to the point that creativity on God's part and man's participation in that creative process become meaningless. In a personal letter, he wrote: "The invisible hand of providence and the action of men, are one and the same thing."[33] World development process is static and unchangeable. Uniformity and regularity usurp sovereignty and creativity in God's ordering of the cosmos. Man's categories again usurp God's prerogatives, and anthropocentrism neglects the dynamism in God's creation.

The work of Karl Barth marks a new theological way of locating the place of man in God's world. The negation of the secular in early Christian and medieval and, to a certain extent, Reformed theology proves as incapable of dealing with this relation of man to cosmos as do the nineteenth-century theologies in their surrender to the naturalistic frame. In Karl Barth, the creative tension between sacred and secular, transcendent and immanent,

God and world, is reaffirmed, making possible serious discussion of man's place in the cosmos.

Barth locates the interrelatedness of God, man, and cosmos in the Word of God. In creation, God has inextricably related himself to the world, and in createdness, man is coordinated to God and world.[34] Man is rooted in the cosmos in lostness *(verloren und verdorben)* but also in foundness and renewal *(wiedergefunden und erneuert).* Although there is no ontology of creation in the Word of God, the Word points to man's responsibility in the cosmos. The lower cosmos *(unterer Kosmos)* is the sphere of man. The dual structure of the world corresponds to the dual structure of man: "The heavens reflect the being and activity of God; the earth, the being and activity of man."[35] Man, linked to God and the world in creation, is teleologically related to God as co-worker. In the realm of history which intervenes, God and man partnership is the character of the relation. God's relationship with the universe is focused in man. As co-worker with God, man, in the ongoing process of subjugation of the earth, participates in the cosmic redemption which is God's will in creation. Barth sees this basic thrust in all Scripture.[36] Responsibility is thus the fulfillment of createdness under the Word of God. Man pursues his destiny as he takes control over the earth as co-worker with God. Barth would place value on this subjugation as man's situation in the cosmos is heightened and God is praised in the growing mastery of mankind over environment.

The thought of Paul Tillich further clarifies the idea of man's responsibility with respect to subjugation of environment. In *Systematic Theology,* he calls attention to Heidegger's distinction between *Zuhandensein* and *Vorhandensein*. While "being at disposal" refers to the technological relationship, "being in existence" denotes the cognitive relationship to reality.[37] Just as man controls his theoretical environment through language, so he also controls the conditions of existence through the making of tools. This organizational capacity of man is part of the givenness of life in its ambiguity.

In Tillich the technical act is always related to the word. Man's environmental control is gained through a logos extension into reality. Reason conceives and theorizes, language articulates, and tools implement control. "Language and techniques enable the mind to set and pursue purposes which transcend the environmental situation."[38] The point of transcendence leads Tillich to discuss the ambiguity of technical mastery over environment. Here the man-environment relationship takes on the dimension of responsibility. Technical transformation of environment, a self-creating function of praxis, bears built-in ambiguity. "The tool which liberates man also subjects him to the rules of its making."[39] Three ambiguities of technical production constitute the tension that engenders responsibility. The ambiguity of "freedom and limitations" of "means and ends," and of "self and thing" characterize all technical production.

The first ambiguity speaks of the possibility, both destructive and creative, that all environmental control brings. Because of man's desire to transcend his finitude, each act of technical mastery is symbolized in a way which seeks to reach the divine sphere. A good illustration of this is the excitement which surrounds the new possibilities in duplicating vital functions in man through artificial organs and prosthetic devices. Chiliasm again emerges in the consciousness of man: the anticipation of a day when human life will be free from imperfection through the gifts of medical science. Responsibility calls man to realize that each extension of technical freedom also carries with it commensurate limitation. The world is far less free since the advent of nuclear weapons. Responsibility is intensified as freedom is thus delimited.

The ambiguity of "means and ends" raises yet another question for responsible man. What is the purpose of his invention? When he exerts new control over environment or informs matter in some new way, what is the designed intention? Tillich notes that technical possibility becomes social and individual temptation. The ability to create somehow inevitably works itself out in implementation. Robert Oppenheimer noted with regard to nuclear capacity that what is technically sweet becomes irresistible.[40] The social dimension of responsibility bears down heavily at this point. Are our national and international priorities responsibly conceived? Is development of more devastating weaponry and space exploration worthy of the priority it takes over the feeding and healing of the poor of the earth? Responsible criteria for expending our resources must be developed commensurate with our technical mastery of environment.

It is the third ambiguity which Tillich locates that is fraught with greatest danger. The ambiguity of "self and thing" refers to the ever-present problem of objectifying something from the natural order, thus destroying its natural structures and relations.[41] As we noted in an earlier chapter, cybernation raises the dual temptation of objectifying the subjective and subjectifying the objective. The danger is that man loses his selfhood among objects with which he cannot communicate. His selfhood is redefined in terms of his ability to produce and direct. "The liberation given to man by technical possibilities turns into enslavement to technical actuality."[42] In the linking of computers to automated machines, this ambiguity is raised. Cybernation can organize human beings out of work; it transforms the remaining technological elite into organization men at best, one-dimensional men at worst.[43]

Tillich claims that the incessant search of man for unambiguous relations of "freedom and limitations," "means and ends," and "self and thing" is ultimately the quest for the Kingdom of God.[44] This quest is itself ambiguous. It is the genius of man's nature as he strives for mastery, but it is also his fall as he succumbs to hubris.

The co-worker motif does not fit schematically into Tillich's thought because of the basic ontological structure. It is purpose rather than design which unites man and God in this task. This thought will be explored in the next chapter. In summary of Tillich's pertinent thought we note that responsibility in terms of man's relation to God as co-worker is twofold. It is fulfilled as realism on man's part accepts finitude without Promethean defiance. Meanwhile, his cybernetic mastery over environment must proceed unhindered along lines of establishing justice and welfare for mankind. This concludes the historical synopsis of the man-world relation with reference to responsibility. We now turn to an analysis of the co-worker motif.

The co-worker relation of man and God in the act of subduing the earth will be explicated and then clarified by criticizing a purely process view of development. A definition of man's responsibility hangs critically on a precise understanding of the co-worker relation.

Standing as the subjugator of the earth man is the expression of God's Lordship in the creation. His character as co-worker therefore is centered in the "ruling over" the world that he claims. The biblical terms are clear in their reference to mastery: כבש רדה to trample on; to tread (grapes), to subdue. The mandate has cosmic significance.

It refers to all matter and energy as well as the animals. Man is stationed in the world to communicate the rational dominion of God over nature and to order reality according to his will.

The co-worker relation is seen in its full intensity only in the context of Jesus Christ. Bonhoeffer states the point succinctly in his explication of the four mandates of creation. With reference to labor he states: "The labor which is instituted in paradise is participation by man in the action of creation *(mitschöpferisches Tun).* By its means there is created a world of things and values which is designed for the glorification and service of Jesus Christ."[45]

Bonhoeffer locates the redemptive theme, which this book considers central to the understanding of cybernation, in two ways. In creation man is given the mandate to participate in the creative, redemptive process of world-building. This task is basic to his createdness. But he also recognizes in Jesus Christ a lost creation and he longs for paradise. With what Tillich calls dreaming innocence man hearkens to that primeval cosmic order and harmony and this prompts his redemptive striving toward new creation. "From the labor which man performs here in fulfillment of the divinely imposed task there arises that likeness of the celestial world by which the man who recognizes Jesus Christ is reminded of the lost paradise."[46] In Christ the cosmic redemption is located. He is the purpose of creation: ". . . all things were created through him and for him. He is before all things and in him all things hold together" (Col. 1:16-17). If we search for the principle of coherence or the object of purpose we ultimately face the question of Christology.

Wolf-Dieter Marsch in an article, "Kybernetik und Ethos," has noted the implication of this point: "The follower of Christ is able to be God's co-worker in the yet unfulfilled creation because the creation is given by God as the realm of mutual work."[47] Marsch goes on to speak of the unfulfilled work of Christ. He speaks of the filling up of the sufferings of Christ (Col. 1:24) and the kenosis passage with its striving toward that day of cosmic fulfillment when every knee shall bow (Phil. 2). Here the redemptive context emphasizes the human dimension although the cosmic dimension, enfolding both animate and inanimate nature, is certainly implicit.

There is the longing for fulfillment as man groans to perceive his destiny as a son of God. There is the agony with which history strains toward fulfillment under judgment and grace. The dimension often overlooked, though ever present in Scripture, is the way that nature herself is part of this cosmic redemption. Man, caught up into Christ, engages both history and environment (matter and energy) in creative striving. In this role he perceives the pathos of this cosmic strife but he is also released in Christ to perceive and foretaste its purpose and fulfillment. And so he knows the truth of Chardin's dictum: "In one manner or the other it still remains true that, even in the view of the mere biologist, the human epic resembles nothing so much as a way of the cross."[48] The cosmic redemptive struggle becomes part of one's being to the man who is caught up into Christ. "Those who hope in Christ," says Moltmann, "can no longer put up with reality the way it is, but begin to suffer under it, to contradict it."[49]

Man as the creature made in God's image alone participates in the *divinus* and *humanum,* the divine sphere and the human. The proper understanding of the responsibility that he has as co-worker is seen only as the duality of his existence is recognized. Like God, he is part of the realm of reason, freedom, and responsibility. His being is rooted in the control and communication dimension of reality. He is stationed a little lower than the angels and crowned with glory and honor. But like all that is created, he himself is subject to change and transformation. He is involved in the travail of nature. He also longs for fulfillment. The responsible man knows the finitude that characterizes his life but he also knows the glorious liberty of the sons of God. He is released in grace to master environment, for this mastery, though fraught with ambiguity, is from God and directed to God. In thoughts reminiscent of those developed in earlier chapters the Christian gospel releases man from the tyranny of bondage to nature and elevates him to his destined station of dominion as co-worker with God.

He is thus free from the compulsion to create either a Promethean or a Faustian situation. When man falls under that kind of compulsion he betrays his responsible co-worker station. "Fill the earth and subdue it," says Helmut Thielicke, can no longer be interpreted to mean "a tremendous glorification of man . . . We are to rule and subdue the earth because we stand under God and are privileged to be his Viceroys *(Statthalter)*. But being a viceroy of the Creator is something different from being a creature who makes of himself a God or at least a superman *(Übermenschen macht)*."[50] Certainly a gracious God ruling a good creation releases man to a responsible role of freedom as his co-worker. But this freedom is creative and not passive as some criticism might imply. Man's responsible posture is not releasing himself to inexorable process of a developing world in some type of mystic absorption. Man must coerce and direct development for good ends. Although the short historical epoch of man's manipulation of the environment is strewn with evidence of his perversity (the great war machines he has built exemplify this), the alternative of passive relaxation into oneness with creation has equally demonic aspects (the Hindu and Muhammadan resignation in the face of inexorable natural process, for example).

Man is called by God not so much to be the builder but to be the organizer. The rich intricacies of nature are challenge and enticement to the theoretical and practical capacities of control and communication given man in creation. To passively disengage in some mystical fashion or to surrender uniqueness and melt into some cosmic oneness is to deny the creation mandate man is given and to deny the co-creative responsibility he has under God.[51]

In the cybernated era man's responsibility moves beyond the level of privilege to the level of necessity. If man is to survive and if life is to be preserved on this planet, man must responsibly pursue his co-creative relation with God in control and communication to inform environment. The global electronic net has already bound mankind into a fabric of interaction and interdependence. What Hans Schmidt has called "the process of mutual effect" is now the basic communicative link between all mankind and in turn all reality. "The preservation and improvement of human life obviously depend on the . . . conscious assuming of technological-scientific responsibility for the world."[52]

The co-worker relation is one that is characterized by ambiguity and fear. To venture into manipulation of environment evokes terror at every step. Nature herself balks and retaliates in response to man's intrusions. Man gains control over disease in some way, then nature regroups and attacks him in a more virulent way. Man gains control over premature death, and population explosion threatens his civilization. He explores the stars while his earthly societies quake with crisis upon crisis. To return to nature after the fashion of Rousseau or even Thoreau is impossible for technological man. He must proceed to dislodge the secrets of nature. He must continue to harness matter and energy via cybernetic mastery. He must subdue the cosmos. Man must move ahead in faith and in compassion. He must trust the Lord of nature to reward his creativity as long as it is done in the name of goodness and humanenness. Man's creation mandate is not merely to control environment with technologic and electronic mastery, but also to fill the earth, to be fruitful, to enrich human existence, to fulfill human destiny in meaningful community.

V

Technology and Hope

> . . . an old French sentence says "God works in moments"—En peu d'heure[s] Dieu labeur [*sic*]. We ask for long life, but tis deep life, or grand moments that signify. Let the measure of time be spiritual not mechanical . . .[1]

"The philosophers have only interpreted the world in various ways," said Karl Marx; "the point however is to change it."[2] In locating man's ability to shape the future Marx deals with a crucial feature of the technological age and man's responsibility. Man's electronic extension within the environment at the planning level gives him a much firmer control over the future. While tomorrow holds only terror for primitive man, technological man can create the tomorrow. When man burst the chains that bound him to earth on April 12, 1961, in Yuri Gagarin, he began to fashion human domain in space. As consequence, he revolutionizes the relationship he has had with his future. Freeman Dyson, of Princeton's Institute for Advanced Study, writes of new hope found in the lunar landing: "Men's tribal instincts will move back from the destructive channels of nationalism, racism and youthful alienation, and find satisfaction in the dangerous life of a frontier society."[3] The future and the frontier as aspects of technological environment move us to the consideration of eschatology as this shapes responsibility.

We have looked at the redemptive motif and the co-worker motif in theology. These both have eschatological dimension. The eschatological category has taken on new emphasis in theology with the work of Ernst Bloch, Jürgen Moltmann, and others in recent years. Technology and cybernation also reflect the hope tendency. In the Marxist-Christian dialogue eschatology becomes an essential ingredient, vitally necessary to the understanding of technological impulse. It is also a central factor in the ethical evaluation of the cybernation process. In this chapter attempt will be made to forward the discussion of the shaping of ethical responsibility within cybernation by showing how responsibility is directed by eschatological considerations. We will begin with a brief historical survey. Reference will be made to the Old and New Testaments and several points in Christian history. The first part is intended to show the way that eschatology has generated technology. Then two sections will show the connection between technology and cybernation as these contrast in Christian and Marxist ideologies. Finally, a responsibility will be analyzed with specific reference to the way it is directed eschatologically.

Jürgen Moltmann has noted in his important writing on hope that man by nature is *in statu viatoris,* in essence dynamic and moving. His response to this dynamism can be positive or negative, expressed as either *praesumptio* or *desperatio.* His existential life as well as his socio-cultural expression reflect the choice between these alternatives. Even this technology can reflect the dual failures of hasty expectation of technological utopia *(praesumptio),* or deprecation of progress and the resultant cloture to the future *(desperatio)*.[4] His eschatological awareness, in other words, shapes his technology as a part of the total way in which he relates to his world.

The Hebrew experience perceived Yahweh as the Creator God who had finished a cosmos that had value and purpose. It was finished, yet it had a "coming" or "becoming" character. Seeing that the creation was good God created man "from the ground" (Gen. 2:7, my translation). With a pliability comparable to the potter's clay (Isa. 45:9), God summons him with a creative future-oriented task: "fill the earth and subdue it" (Gen. 1:28). The same plasticity marks the creation with respect to man's subjugation of nature. It is open to the future. We have noted the redemptive theme in the Old Testament. The creation is going somewhere and man is a part of the dynamic of that process. This is confirmed in that he stands under the creation mandate.

Although the primitive world throbbed with energy that challenged man's control, it is very difficult to document the notion that early notions of technology were motivated eschatologically. Arend Van Leeuwen notes, however, in his important book[5] the way that technological dynamism is rooted, not only in Western Christianity, but in its Jewish antecedents. The Hebrew enterprises of temple- and nation-building were surely motivated by hope as much as by command. Beyond the concrete expressions, technological impulse is given profound direction by two Hebrew concepts. The first, already mentioned, is that man is under creation mandate to subdue the earth. The second is more significant. Man is created *imago dei* (Gen. 1:27; 9:6). The distinction that man is given in the creation has considerable import in directing man's emergence as a creator of technologic and cybernetic control of environment. Ernst Benz says that the *imago dei* "has become one of the strongest impulses for man's technological development and realization."[6]

In the New Testament eschatological factors emerge that are very directly related to technology. The metaphors used by Paul are frequently technological images. This is seen when the Apostle speaks of man's mutual work with God. The technological images of the building and the field control sections of the Corinthian literature (1 Cor. 3:9-10). Although the controlling theme here is kingdom-building, the point stands that the relationship between God and man is given eschatological dimension. Man is commissioned to be a skilled master builder, building on that foundation which stabilizes and controls the design: Jesus Christ (1 Cor. 3:11).

The dominion motif again calls attention to the stature of man's work in the creation. The astronauts recalled Psalm 8 with awesome praise that God has given men dominion over "the works of [his] hands." Only in this perception could they deeply convey what their feat was all about. When man conceives of himself in a way that links him to the Creator in a relationship of mutuality and common purpose, and when that unity is found in dominion over the world, a powerful impulse is released that generates high technological accomplishment. Theological writing at the time of the Renaissance, the Reformation, and the Enlightenment strongly stresses this dominion theme as it encourages human creativity. Man feels compelled to wrestle order and beauty from the environment when he believes that he shares the informing and creative work of God. Hans Lilje and Paul Tillich have shown this point in different ways.[7] Frequently this press toward the future has taken the form of striving for a technological utopia. The co-worker relation has often been denied and a view has been substituted that sees man as "the measure of all things." The future is frequently seen as the creation of man or an inevitable working out of the progressivism inherent in reality. Yet often at the heart of the technological impulse has been the conviction, first felt in the New Testament, that in Jesus Christ the clue is found to the direction in which the creation is moving, indeed that in him, the entire organizational unity of creation (both animate and inanimate), coheres and is directed (Eph. 1:10). The eschatology of the concept is clear. As expressed in Colossians all future design, direction, and consummation of creation inhere in the person of Jesus Christ. The word ἀνακεφαλαιώσασθαι, though literally referring to a presently existing human network, has clear reference in the Ephesian passage to a trans-temporal cosmic network of organization. The subduing of the cosmos is an advent transformation of reality redemptively located in God who shares cosmic dominion instrumentally with man.

The concept of time implicit in the New Testament also contributes to the eschatological dimension. The view of time fundamental to the New Testament is linear, not cyclic. Although history is the stage of an ever-present struggle, a conflict between good and evil forces, it remains in transit to final purpose. Time takes new meaning and significance since it is the arena of God's activity. To be God's fellow-worker in this context brings with it an urgency and a high responsibility. Since God has measured time in his purpose, the man of God is under compulsion. This compulsion is directed by the end or purpose which presses in on the present moment. Time becomes for man a measure that demands responsibility, that is directed eschatologically. "We must work . . . while it is day" (John 9:4), "making the most of the time" (Eph. 5:16).

Benz has pointed to the way this New Testament eschatology has injected a sense of progress and acceleration into the technological life of the Christian West: "This new concept of time also led to an immeasurable acceleration of technical inventions and of technological solutions of the practical problems of life."[8]

The first Christian theologian to discuss technology was St. Augustine. A glimpse into his thought as it contributes to the understanding of the eschatological directing of responsibility will be of value at this point. The eschatological note prevails at the end of the magnificent section of Book XXII of *The City of God.* God's initial work in creation is in no sense a closed action. He works presently and will work into the future (John 5:17). His spirit is the sustaining power giving coherence and direction to all reality. ". . . if God withdrew, even from inanimate things, His creative power, they could not continue to be what they became by creation, let alone complete that series of movements which were meant to measure the span of their existence."[9] Augustine goes on to show the way that divine power energizes human creativity, principally in procreation, but also in the cultural creativity of the arts, communication and other forms of technology. Man in his creative genius, which is rooted in the *imago dei,* is able to rationally perceive reality. He can show ingenuity and artistry; his mind can fashion argumentation and subtle precision in communication; all this is the gift of God to the natural man. Here man shares in God's power and his nature. (Augustine notes that even man's posture points to this uniqueness: "Man was made to walk erect with his eyes on heaven, as though to remind him to keep his thoughts on things above.")[10]

The predestination theme also contributes to the eschatological dimension of man's responsibility. All of the marvelous workings of nature ("The grandiose spectacle of the open sea, clothing and reclothing itself in dresses of changing shades of green and purple and blue")[11] are governed and directed by the providential action of God. The life of man in all the dimensions of his creativity is also ordered by God. At one point he envisions with Pauline insight the marvelous things to come from God "who did not spare his own Son" (Rom. 8:32). "Remember, all these favors taken together are but fragmentary solace allowed us in a life condemned to misery."[12] The full riches of human creativity will be known only when the elect see God face to face. "Just imagine . . . how strong will be the human spirit when there will be no passion to play the tyrant or the conqueror . . . think of the mind's universal knowledge in that condition where we shall drink, in all felicity and ease, of God's own wisdom at the very source."[13] This kind of expectancy is what prompts man to plumb the depths of creativity in this life. Man senses dimensions of control and communication that are residual or unexploited in his nature, and he strives to call these gifts to the surface. "Hope," said Kierkegaard, "is a passion for the possible."

The lasting contribution of Augustine to our theme is found in the way he sets present creativity in both a creation and a futuristic setting. Man responsibly engages in technologic or cybernetic mastery of environment as he acknowledges that his life is situated in the creative action of God; as he recognizes the fact that his ingenuity is rooted in his createdness in the image of God; as he perceives that world order and change are dynamically given by the perpetual impulse of the divine Spirit. When man's life is eschatologically shaped in the way mentioned he is responsible. His own creative activity is not pioneering in the sense that he works alone. He is co-worker as divine power penetrates his conceptualization, decision-making, and implementation. Finally he is responsible when he humbly acknowledges the conditionedness and limitations of his creativity and of the things he has made, when he anticipates the fullness of cosmic beauty and order that is perfected only in eternity.

The polarization of secular and sacred, the city of man and the City of God, remains a conceptual problem in Augustine. This theme, however, does not fall within the scope of this book nor does it detract from the significant contribution his thought makes to this thesis.

The monastic movement in the West contributed to the intensification of the eschatological compulsion that directed the development of technology. The Benedictine and Cluniac reform movements in particular highlighted the *ora et labora* principle. Stress was given to the urgency of the time expended even in the arts *mechanicae* and the *artificia.* All of these enterprises were to be carried out with great skill and care because of the eschatological press.[14] The agricultural technology that developed from the monastic emphasis was formidable. An eschatological dimension was heightened in this period that would be emphasized again in the rise of capitalism in the West. Simply stated the principle claims that technological activity was a means of overcoming original sin. Since man's work originated in the Fall the compulsion to subjugate environment became, not so much an attempt to overcome the Fall, but rather to work out its consequences. If man has to wrest his livelihood from the environment because of the fact that the environment after the Fall is antipathetic to his desires, one sees clearly how this becomes technological impulse. In this mood Shiller spoke of the way in which the elements themselves hate what is created by human hands. Not only for his own survival, but for the integrity of his life also, man must incessantly strive to control his environment. His destiny is somehow caught up in the way he draws order from the environment.

We could make reference at this point to the origins of the industrial revolution. We could refer to the eschatological impulse of the pietism of the eighteenth century in Germany or the desire to build the Kingdom of God on the earth that arose in the next century. Suffice it to say that the eschatological dimension forcefully colored industrial development as it took shape on the Continent, Great Britain, and the New World. The subject is difficult to deal with in the context of this book because of the way that secular eschatologies intermingle with theological eschatologies from the Renaissance onward. A more fruitful approach would be to now move from the historical to the analytical approach. The major eschatological themes have been explored in historical setting. All through the ages of technological development we see the thread of eschatology weaving its way. The Christian understanding of the nature and destiny of man and the world have irrevocably shaped the rise of technology.[15]

We have been speaking mainly of eschatology with reference to technology in general. Now we must show how cybernation is connected to technology, indeed how it grows out of and extends technology. Cybernation can be called the extension of technology because when analyzed eschatologically it becomes certain that only belief in design and predictable future generates power and will to extend control and communication into the environment cybernetically. In fact, only an uncanny confidence in the rationality of nature and a rather childlike hope in the fruitful consequence of action could possibly direct cybernation. To understand this phenomenon the only other alternative would be to posit complete irresponsibility on the part of man, irresponsibility which manipulates environment only for the sake of the desired pragmatic result, with no thought given to the consequences or the implications. The former alternative is supported by the thought of Marshall McLuhan in his book *Understanding Media: The Extensions of Man,*[16] and of Buckminster Fuller in his visionary writing.[17] The thesis of each of these men is that a confidence in the predictability, rationality, and continuity of nature is a principle factor in the motivation that has led man to electronically extend himself in the subjugation of his environment.

Cybernation then, as we have defined it in this book as the duplication of human function with electronic circuitry, is but the extension of that technological impulse that has always been directed eschatologically. When man extends control and communication into the environment he extends his nervous system, with its facilities for decision-making, analysis and synthesis, and feedback evaluation. This is an extension of the same order as the extension of his manual facility through the machine. Because of the high order in which reality is penetrated in cybernation, responsibility is heightened. The eschatological dimension of responsibility is relevant because man in his cybernetic capability is able not only to predict but to shape the future. Goals and directed action are informing principles of all cybernetic activity. The location of these goals and the selection of means to achieve them are decisions of high order responsibility.

To discuss the eschatological direction of responsibility we must now examine two ideologies—Christianity and Marxism—as these conceive of reality in an eschatological way and as these direct cybernation, particularly with reference to responsibility. What is right and legitimate and what is undesirable in each system? What are the goals and means of implementation and how can these be ethically evaluated? The cleavage and convergence of these ideological positions provide the major direction by which practical decisions are made every day in this technological world.

One focus of the Marxist-Christian dialogue has been the question of knowledge of the transcendent. Also current is the discussion of the social ethic that results from the answer to that question. The classic Marxist critique of religion still hovers in the background of the intellectual dialogue, asserting that the man and society that postulate God or transcendent power that has control in history is only escaping responsibility. This argument of Marxism is certainly well documented in history and the evidence to support it is devastating. A theistic world view has often been used to rationalize worldly irresponsibility. The man, to paraphrase George Bernard Shaw, whose God is in the sky, often makes hell upon the earth. Man often erects the technological structure because vertical environment control comes easier than the horizontal. Many critics of the space program from the black poverty community have, for example, noted our power at the upward thrust of technology contrasted to our failure to use our ingenuity and skill toward the solution of human problems.

When true to itself the Christian faith affirms the eschatological character of man's being and activity in the world. Only a man who lives in a world with God can meet the future with optimism and hope. Although the Marxist mind sees meaningful futuristic planning contingent on an open world *(offene Welt)*,[18] the Christian argues that it is only in a world view where God is known as the one who holds the future that that future becomes open to creative planning. The man who is released to the future through such hope can responsibly shape that future in a meaningful way. It is precisely because of the futureness *(Zukunftscharakter)* of man's life that he can claim cybernetic mastery over the environment.

Gerhard Ebeling has expressed a word of caution at this point. Man can be tempted to surrender his freedom and futureness in cybernation by accepting an inevitability as future planning becomes more thoroughgoing.[19] Although Bastian calls this an anti-cybernetic protest *(antikybernätisches Einspruch)*,[20] Ebeling has a point when he contends that individuality and the correlative future orientation of man is threatened by irresponsible collective planning. The question stands whether man's *Zukunftscharakter* is enhanced or ameliorated in cybernation. The fear that Ebeling expresses would be more appropriate were the Christian to foresake his responsibility in control and communication of the environment leaving the shaping of the future to those whose only motive is expediency or aggrandizement.

A fundamental difference in definition is at work here. The Marxist ethos defines the fulfillment of man in terms of freedom and universality. When man is released from self-striving to affirm his communal solidarity, he is a man. Ebeling, standing in the Christian tradition, sees more importance in individual responsibility. The people who know God to be sovereign in history perceive with hope a future that is meaningful. They seek a tomorrow that demands responsibility and seek to creatively shape a future that accords with the divine will, insofar as this is humanly possible to know.

In Marxism a solidarity of consciousness between man and the world is affirmed. Marx developed his exposition on the freedom of the individual and of the relation of man to nature in his early writing.[21] In the active relation man has with nature, indeed in the vitality of the interaction, is found the dialectical reality that constitutes his essence. In this perceptive intercourse man comprehends nature and is able to constrain it in vitalistic ways into the future. Man has his "free being" and "universal essence" as he plans and acts according to dialectical impulse which is the Marxist eschatology. Man's uniqueness, says Marx, exists "in practice precisely in that universality which makes all nature his organic body . . . man knows how to produce in accordance with every genus, knows how to apply the criterion proper to each object; that is, man produces in accordance with the laws of beauty."[22] Commenting on this passage, a sympathetic Lindsay says that within this framework man alone can carry on a sustained, active relation to nature, with roots both in past and future. "Man alone can plan and direct his acts in a comprehensively realized plan."[23]

Action as well as knowledge is shaped eschatologically. Here at the point of value formation clear contrast is also seen. What is the good according to Marxism? What ought I to do? Here again the basic metaphysic with its realized eschatological character comes to bear. I ought to do that which enhances freedom and obtains universality. The responsible man is the man who participates with such involvement in nature that he shares in potentiality as well as actuality. Through creative engagement in the dialectical process, man in a sense creates his own future, which is good. The good is that which hastens the dialectic, that which ferments the situation so as to create change. This is not simplistic progressivism. The Jewish apocalyptic note is felt in Marx, as Toynbee notes,[24] so that progress is not an inevitable pleasant process of betterment. There is anguish and conflict at the heart of historical process. The good, in short, is that activity, that planning, which seeks the organic wholeness of man with nature. The industrialization process has a profound dehumanization tendency, of course, yet it is in the economic sphere that the good is most profoundly realizable. The good, according to Marxism, is that which facilitates the social process. With reference to the uniqueness of man the good is defined as personality formation as this takes place principally through the productive facility. Production here is used in the comprehensive sense to refer to the total creative enterprise of the human personality, but with special reference to the economic sphere. Production is the good because it articulates the highest need of man. A contemporary Marxist has noted that "... production is the necessary spearhead of the human need to master nature."[25] Engels has commented in a similar vein that it is in "... changing [external nature], making it serve his ends, he dominates it."[26] The good is located at the juncture where man manipulates the environment. It is a reality that is present when the spiritual is firmly united with the material. Goodness is not found in being or action alone. The two must intermingle in creative worldly activity to produce the good. Man, in fact, cannot be spoken of in abstraction. He exists only in the action arena, in this engagement in history, this solidarity with society.

The Marxist ethic has then to do with methodology and historical dynamics rather than structure and content. One acts ethically as a goal is pursued and achieved. The valuative act is in the making of the goals, the means are evaluated only in terms of the goals.[27]

The critique that Christianity brings bears on this point. When the goals of technological progress become the objects of ethical evaluation rather than the means or the subjects involved, man is reduced to a functionaire.

Man's value is measured by his pragmatic worth *(Funktionswert).* Although it is an eschatological picture of man that we find in Marxism, it is a defaced picture. In Marxism the "picture of a particularized eschatologically conceived man becomes vague and indiscernible. His individual character is lost as he is assimilated into the bland collective consciousness."[28] In this critique of Marxism, Thielicke's main point concerns the way that man is reduced in this ideology to the state of the functionaire (though this is disguised in the discussion of the collective consciousness), thus surrendering the dignity that his life has because of his relational and eschatological character.

The Christian understanding of cybernation and technology allows man to rejoice in mastery over the environment for the simple reason that this facility enables the extension of the human genius. Man's dignity is not found in his function but rather in his eschatological relatedness.

Ethical evaluation in the Christian understanding comes at this deeper level. It is not the technological goal or even the means to that goal that are the objects of ethical focus, although they certainly are ethical considerations. The focus of the Christian ethic is the responsible action of each man as he stands under the judgment and grace of God, as his decisions qualitatively conform to the will of God in its personal and communal dimension. Valuation cannot be made solely with reference to the technological act. Behind each act is a way of life, a philosophy of history, a formed conscience, expressed in a world view which is only reflected in the technological or cybernetic act. Ethical responsibility is located in the *Weltanschauung* and resultant way of life rather than in each technological act. The responsibility for a war machine or an automated system that grinds human beings in its production is ethically vulnerable but so is the human society that produces, sustains, and condones such a system. The fabric of guilt is complex with many factors woven together. The Christian understands this depth meaning of ethical responsibility as expression of the Fall. The central responsibility of the Christian is not then to scrutinize each moment of technological advance separately but rather to examine the whole fabric of social purpose. It is useless, for example, to criticize the use of an artificial heart or the technique of heart transplantation on ethical grounds when the society has already decided that length of life is a primary value. The responsibility of man with specific reference to cybernation as with all activity has to do not only with the directive of conscience, but also with the formation of conscience. We must discuss at this point the view of conscience current in the divergent ideologies with specific reference to the way that conscience is formed.

It is an eschatological compulsion that shapes conscience in both the Marxist and Christian system. Yet the future or "comingness" of reality is conceived of in different ways. The future in Marxism has an inherent, built-in inevitability; the future in the Christian cosmology is impending. It has an advent character.[29] In Marxism conscience is defined in its collectivity. Since the view of conscience always emerges out of an anthropology we must set it against this backdrop. In Marxism, as we have noted earlier, man's identity is wrapped up with his solidarity in community and within the historical process. Man in his essential nature belongs to the environment out of which he emerges. Organically he rises from the natural world and historically he stands in a certain *Zeitgeist.* His conscience is shaped according to the apocalyptic anthropology of the system. History is in a convulsive, cataclysmic state, straining incessantly toward that idyllic future of the classless society. The future is indelibly potential in reality. To say that man's conscience is shaped apocalyptically is a refinement of what we have called the eschatological direction. Conscientious man strains toward that future, participates in that convulsion, and, through all phases of his cultural enterprise (including his technology and cybernation), struggles to bring the future into the present.

It is difficult to speak of conscience as such in the Marxist system. Although the literature has a strong moralistic flavor,[30] the subject of morality is defined in a transpersonal, collective way. The individual who reaches for universality and freedom is moral in that activity. Those activities which thwart such development are evil. What then is the moral arbiter? There are the previously stated absolute principles which are intrinsic to nature and human nature. Here Marx is a casuist in that he sees these principles as inviolable because of the way they have been validated in the historic development of man in society. They are not given or inherent; rather they are emergent values in the human situation which have lasting verity for that reason. The relativity of values is a stronger note in the Marxist tradition. Hegel claimed: "The deeds of Great Men, of the World-historical Personalities ... must not be brought into collision with irrelevant moral claims. The litany of private virtues, of modesty, humility, philanthropy, and forebearance [*sic*], must not be raised against them. The History of the World can, in principle, entirely ignore the circle within which morality . . . lies."[31] Stronger in Marx than Hegel is the stress on cultural formation of conscience, rendering relativity as the norm. When the dynamism of dialectical movement is frustrated, ethical principles, regardless of their value, must be suspended. Conscience is that point of consciousness where the free spirit of man is informed by the noblest of the cultural residue. Lindsay succinctly states: "The stable ethic resides in all that in a given society which makes for cooperation, union, freedom, and effective control of nature."[32]

The Marxist view of conscience thus can be finally analyzed in a cyclical manner. What the free and universal consciousness perceives must be planned for and implemented. What is needed must be developed. The futuristic goals that are socially desirable must be fed back and reinforced to inform the emerging moral conscience of the society. Here we approach the new world that has been gravely forecast by Huxley and others, a world going through what he called the "final revolution" where technology is applied to human affairs and man himself is manipulated through "technicization" or psychobiological control.[33]

The Christian view of conscience calls the Marxist view into question because of the relativity of judgment which characterizes that system. If man is left to the tyranny of subjectivism or the "least common denominator" ethic of consensus, his situation is grave. David Riesman in *The Lonely Crowd* has spoken of the bankruptcy of the "outer-directed person" whose values are shaped totally by the expectations of others. Strunz likewise has called into question the *"kreiselkompaß"* orientation in decision-making, along with other views of conscience.[34] The Christian view of conscience is not one which normalizes the prevailing situation. A deeper view of Christian conscience, one which is rooted in the eschatological dimension of faith, alone can provide meaningful direction in a cybernated age.

Recognizing the anthropological setting of conscience, Thielicke explicates the relational view of conscience which is at the heart of the Christian tradition.

The conscience is sensitized as it relates to another. The other may be interpreted as the voice of God or as practical reason. It is in the tension of the confrontation of man with himself that the conscience awakens.[35] The cleavage of human existence is focused in the conscience. Here is the focus of the tension between the *ought* and the *can*, between the imperative and the indicative.

In the opening stanzas of the Letter to the Romans, Paul speaks of the futility that claims the lives of those who change the truth of God into a lie and worship and serve what is created (τῇ κτίσει) rather than the Creator (τὸν κτίσαντα) (Rom. 1:25). The conscience (συνειδήσεως) witnesses to man's heart at this threshold of tension concerning the disparity of his present life as over against the law of God (Rom. 2:15). A false object of worship, a conditional reality raised to unconditional status, is fundamental to the failure of man. The elevation and subsequent adoration of the creature (something in the natural order) expresses the fall of man. Yet man is justified by faith (Rom. 5:1). Faith ushers in a new perspective on reality, one which not only endures the present but is animated by a lively hope (Rom. 5:2, 4-5). The man who abides in Christ lives in hope. Only this man is released to the future because he alone has been freed from the compulsive concentration on the present and past which is sin. Conscience, released from dread, is thus freed to a measure of discernment, a means of sensitive perception of the future that provides guidance for responsible planning. Conscience in the redeemed state ceases to function solely as the condemning judge.

Faith in this context becomes a venture. Thielicke from his pervasive eschatological understanding of conscience calls it "... a flight from myself ... toward the great possibilities of God."[36]

We have looked at two of the questions that Immanuel Kant posited as basic to man, "What can I know?" and "What ought I to do?"[37] With reference again to the contrasting world views of Marxism and Christianity we now look at his third question, the question of hope. What future can man hope for? Of particular interest to this inquiry is the way in which this anticipation shapes responsibility. The Marxist and the Christian answer the question "What can I hope?" in ways that converge at some points and diverge at others.

Before attempting to answer this question a preliminary question must be faced. This question concerns the source from which the future comes. Here the basic world views of Marxism and Christianity contrast. The contrast is most pointed with respect to the way that eschatological forces direct the future. Moltmann has pointed out the way in which Christian eschatology is directed by both unveiling and fulfillment. Not only is a new creation being unfolded to man but a promise is being fulfilled.[38] The givenness *(Ergangenheit)* character of the promise, contrasted to its pastness *(Vergangenheit)*, thrusts the hope of man toward the future. The Christian man moves with the future contingent on the future of Jesus Christ.

The Apostle Paul shows how the destiny of man is intertwined with the destiny of the world as the whole creation strains for that future fulfillment (Rom. 8:19 ff.). All of the creation waits in hope for the glorious fulfillment in Jesus Christ. All nature thus is striving toward its future which is held by and shall be consummated in Jesus Christ (Rom. 8:23; Col. 1:16, 20).

Sympathetic to Christian eschatology is the secular humanism ably represented by the thought of Ernst Bloch. Bloch contends that the key that opens to the understanding the meaning of human existence is found in the hope that he has for the future of humanity and for the world. Although poignantly longing for a new world and new humanity, the Giver of that future is historicized in Bloch. Here we have a secular messianism. Philosophically Bloch builds his thought on an ontology characterized by the phrase "NOT YET BEING" *(noch nicht seins).*[39] The key distinction found in this tradition is that the future is humanistically directed. In images strikingly similar to the biblical images of the Kingdom of God, the Marxist anticipates "... the transcendent homeland where all who now suffer, labor and are now incomplete will find their true identity."[40] Although Bloch has become a pivotal point of dialogue with the Marxist system, his thought only clarifies the radically different source from which hope proceeds in the system.[41]

The Marxist answers the question "What may I hope for?" with a clear-cut apocalyptic answer. Man can hope for a new social order; one which will be free from the tyranny of class structure and the resultant exploitation. History is inevitably bringing in this future. Man, through his effort, planning, and in some contexts revolution, can hasten the coming of this desired world. This new world is contingent on technical and social development. Indeed the full focus of human responsibility centers here. The Marxist sees the responsibility of man as well as the eschatological substance that directs that responsibility located in the historical continuum. Leszek Kolakowski, a Polish philospher, points this up in his explication of *Die Hoffnung und die Historische Materei*: "that a scrupulous conscience only has power if it sees itself as part of the historical process."[42]

Here we see the reason for the enthusiasm with which the Soviet Union and the other nations motivated by Marxism have greeted the advent of cybernation. Man's responsibility is shaped in this historical-social continuum which is self-contained. Therefore, planning—indeed all dimensions of cybernetic mastery of the environment, particularly the human environment—is welcomed. When not only technology but the human enterprise is seen as malleable to the programming of the state we see why the development of cybernation is such an important development in Russia. The endeavors of bio-medicine, electric technology, and space research reflect the Soviet philosophy at this point. The American ethos exults in an all-American boy stepping down on the moon. In the Soviet Union technological feat overrides the importance of the human factor.

The question of what the future should be is also clear in Marxism. In the same way that Hegel noted that man is exonerated by the machine, the Marxist sees in the cybernation process the liberation of man. Although the economic struggle for the classless society centers in the technological arena, the cessation of the struggle and the eventual utopian resolution will be found in the society that transcends the wage-earning struggle. The productive necessity, which, as we have noted is intrinsic to the being of man, can then emerge in its fullness as man utilizes the full powers of his creativity in culture.

In the Marxist system the limits of responsibility are defined by the socio-historical context. Man's responsibility receives no higher direction than the immanent social process. In words that are non-Marxist in origin, the viewpoint of the ideology is clearly expressed: "... to be a man ... is precisely to be responsible ... It is to feel, when setting one's stone that one is contributing to the building of the world."[43] Antoine de Saint-Exupéry in this passage from *Terre des Hommes,* along with Chardin's *Building the Earth,* shows the closeness of the Roman Catholic view with this productive definition of responsibility which controls Marxist thought. The convergence of progressive Christian and Marxist views of man's cosmic responsibility seems to be a crucial component in our desires for co-existence and world peace in the world today.

There is, however, danger in this view of the future. In a society where men are valued because of their productive capacity or their function, inhuman tyranny can result when man, conceived functionally, ceases to serve his function. Creativity in such a future society would be limited because of the fact that the richness of human personality would be channeled into purely pragmatic directions. The horror of the totalitarian state, depicted by Orwell and Huxley, looms in the future of even the most sophisticated cybernated society if the value of a man is thus reduced.

The Christian hopes and works for a world that is in many ways similar to the Marxist hope, but fundamentally different. In general, technological impulse in both East and West pursues the goal of a fuller life for every human being and a more meaningful society. Although the primary hope for the Christian is trans-cosmic, i.e., it transcends the limited hope for a perfected earthly society, he has a definite hope and responsibility for the world. The biblical command to love the neighbor, to execute justice in the earth, carry great responsibility which direct his most sophisticated technological venture.

It is through the cybernetic mechanism—through control, communication, and feedback—that man can best implement his striving for a fuller human life. The Church and Society Conference of the World Council of Churches in 1966 recognized this:

> ... often certain valuable goals come in conflict with others which are equally valuable ... Here the principle of feedback, so central to the modern scientific and technological enterprise, suggests an

> indispensable element in goal-setting. Feedback means the continuous adjustment of processes in the light of their effects. We suggest that feedback . . . must take into account the human elements within which technology functions.[44]

Cybernation here is a liberating force that can be a vehicle for the pursuit of the Christian purpose for man in society. A German engineer, Klaus Tüchel, has shown how humanization in the life of man is possible as cybernation enables the overcoming of specialization and the emergence of cooperative endeavor *(Zusammenarbeit).*[45] In a sense this basic consideration overrides the ideological contrast. A basic need in contemporary life is for man to surrender ideological dogmatism which inhibits genuine universal human concern. The Geneva Conference stressed the necessity for all men whether religious or secularist, Marxist or Christian, to work together with full strength for peace in the emergent world today.[46]

Beyond the common search for universal humanitarianism and social forms that are conducive to such, the Christian faith has a peculiar hope for a future earthly society that is very much shaped by the technological enterprise of man. Exploring this theme briefly with reference to the early modern thinker Walter Rauschenbusch, and then to the contemporary thinker Paul Tillich, will illumine the question of what kind of future the Christian hopes for and how his technologic-cybernetic programming appropriately participates in that future.

Walter Rauschenbusch stands at the turn of the century and clearly evidences a brilliance of social concern that was built on the conviction that the Kingdom of God on earth was the essential message of Jesus.[47] Rauschenbusch was not, as his critics have sometimes asserted, a believer in earthly perfection. He knew that here there is no lasting city. He had, in fact, had a millennialist-like apocalypticalism. "The astronomical clock is already clicking which will ring in the end." Yet he passionately pursued the Master's prayer, "Thy kingdom come . . . on earth."[48] Although evolutionary optimism pervaded the time with its naïve song "onward and upward forever," Rauschenbusch brought a realism to this mood. "In asking for faith in the possibility of a new social order, we ask for no Utopian delusion. We know well that there is no perfection for man in this life."[49] But all the while the Kingdom of God in its earthly dimension is anticipated. It is being unfolded as the church moves through history. "Since the reformation began to free the mind . . . there has been a perceptible increase in speed. Humanity is gaining in elasticity and in capacity for change, and every gain in general intelligence, in organizational capacity . . . increases the ability to advance without disasterous consequences. The swiftness of evolution in our own country proves the immense latent perfectability in human nature."[50]

Rauschenbusch, despite the naïveté of his historical analysis, anticipated a future society that would blossom in history as the fruit of Christianity. In this new world injustice would be ameliorated and society would be humanized. In the new world man would be released from the tyranny of the sweat shops, the agony of child labor, the indignity and dehumanization that the machine and factory—indeed that the whole industrial revolution—had wrought, and the vicious relations between persons that are unleashed in unrestrained capitalism. The new world is a society that is redeemed by the permeating spirit of the redeemed people of God into a new community of human relatedness. For Rauschenbusch the Kingdom of God is this redeemed social order. Although it has both individual and social dimensions it remains personalistic. There is no great emphasis on the humanization or control of nature. "The kingdom of God is humanity organized according to the will of God."[51]

His eschatology is both futuristic and realized. The latter understanding is stronger. "Christ is immanent in humanity."[52] The eschatological dynamic is immanent in the historical community of faith. Here Jesus Christ is giving momentum to human life through the church. The new world that is anticipated is characterized by its faith-relatedness to God and the resultant interpersonal and social renewal that accompanies such faith. There is little discussion of renewed nature or redemptively ordered environment. A similar theological mood present in Rauschenbusch's contemporary Harnack is mediated to the modern discussion in Paul Tillich.

Tillich, at the end of his *Systematic Theology,* speaks of the end of history as the elevation of the temporal into eternity. "History," he says, "is creative of the qualitatively new and runs toward the ultimately new."[53] This new is never completely realized because the ultimate always transcends the temporal moment. Tillich speaks of the content of that life that is eternal, symbolized by the Kingdom of God. After criticizing the ineffable, inexpressible mystery concept of the Kingdom and the continuance theory which says that the Kingdom is basically a continuity with present existence, he speaks of the dynamic, paradoxical understanding of the Kingdom of God. In this Kingdom the "present end of history elevates the positive content of history into eternity at the same time that it excludes the negative from participation. . . . nothing that is created in history is lost, but it is liberated from the negative element with which it is entangled within existence. What happens in time and space, in the smallest particle of matter as well as in the greatest personality, is significant for the eternal life."[54]

Tillich, here with strong historical substance, speaks of eschatology in cosmic terms. The fullness of the creative enterprise in history—contrasted to the pessimism of Bertrand Russell, for example, who contends that all the genius of human creativity is destined to extinction in the vast death of the solar system—is caught up into eternity, contends Tillich. Not that all is eternal but all has the scrutiny of eternity because all that is created strives eschatologically in freedom and destiny.

Man's technical progress, his cybernetic mastery of environment, is fraught with ambiguity. Not only is this enterprise discerned but it undergoes divine cleavage under the eschatological thrust of the Spirit. As man imbues his technical creations with subjective qualities, "under the impact of the Spiritual Presence, even technical processes can become theonomous and the split between the subject and the object of technical activity can be overcome."[55] In the Spirit nothing can merely be a thing—it is a bearer of form and meaning and therefore has eschatological character. If man in his cybernetic mastery of environment can maintain this creative relation with the artifact with which he extends his facilities into the environment, he shall avert the danger of his technology mastering him. Only as man sees his cybernetic mastery under eschatological impulse is he fully responsible in this age.

A good example to illustrate the ethical possibility of cybernation directed eschatologically is a hypothetical consideration regarding the use of the atomic bomb. If proper feedback were available and heeded and if an eschatological design for the future were current both with the dealer and the recipient of Hiroshima and Nagasaki, the travesty would not have been necessary. If communication remained open the disaster might have been averted. If Japanese leadership could have known beforehand the impact of the devastation of the bombs or if American power could have been asserted in potentiality rather than actuality, which is the only sensible way to assert strength today, a black, inhuman day of history might have been averted. As the Geneva Report notes: "It is only as the flow of information is undistorted that men and nations can seek the various goals within a common commitment to human fulfillment. Technology requires a capacity to envision the future and to make explicit what we want and expect from it."[56]

As we search for the solutions for the great problems that face mankind—world hunger, population pressure and explosion, economic transition—man must employ careful planning and prediction. He must use the cybernetic wisdom that he has been entrusted with. The future must be forecast and planned responsibly in the light of the best eschatological direction that is available to him.

In the light of an eschatological awareness of the possibilities open to man in the cybernated era we must say that the responsible man sees himself under God as both master and steward of the world where God has placed his life. The God who has given the world in its fullness is the same God who has ordered that world in ways that are discernible to the human intellect and responsive to his control. He has also set man upon the earth to govern it with responsible concern for the future. Man can transcend the present. He can reflect on the past, and he can contemplate the future. Indeed in the cybernated capacity he can foresee the consequence of certain actions. This man must realize that he labors in freedom under the

Lordship of God, who desires that the future be fulfilling and good for all his creatures.

We cannot believe that God wills that his creation be as it now is: the agony of humanly inflicted pain, of the wanton violation of the natural beauty and resources of the earth, of human strife of every kind, of the ravages that natural disaster still inflicts upon man. All of these dimensions of existence are discordant with the divine will. Although they remain inscrutable, and although it remains true as Chardin has said that even from the view of a scientist the human epoch resembles nothing so much as the way of a cross, God beckons man toward the future. He calls him to use the full creative powers of his intellect to tackle the besetting ambiguities of existence. He calls him to use discernment, decision, and action to chart the course of the future with responsibility. In this context, which we have argued is eschatological, man is the "keeper and transformer"[57] rather than the conqueror.

We have attempted to deal with the eschatological dimension that brings ethical direction to cybernation. We have surveyed the historical connection between hope and technology. We have contrasted the Marxist and Christian anticipation of the future as this shapes cybernation. In conclusion we have seen that responsibility in this era has the twofold character of stewardship and mastery under the Lordship of Jesus Christ. Our final task remains: How is responsibility actuated within the cybernated era?

VI

A Reconsideration of Labor and Leisure

Imagination, not invention, is the supreme master of art as of life.[1]

Cybernation offers man the option to "tune in" or "tune out," to engage or disengage, to involve or isolate, to labor or leisure. The great ethical questions of the electronic age are shaped in this dialectic. At the dawn of the electronic age, Norbert Wiener, the father of cybernation, said: ". . . I have been delighted to see that awareness on the part of a great many of those present [representatives of business management] of the social dangers of our new technology and the social obligations of those responsible for management to see that the new modalities are used for the benefit of man, for increasing his leisure and enriching his spiritual life . . ."[2] It remains to be seen whether this optimism, so out of place for Wiener, shall genuinely happen. Cybernation, as we have noted, certainly has the possibility of liberating man to the genuinely human activity of glorifying and enjoying God *(The Shorter Catechism),* which is, in the Sabbatarian-like understanding, a leisure activity. The man who is released (technically unemployed) from bondage to the production cycle has greater capacities of human freedom. Cybernation also has the contrary tendency as we have noted in Chapter III. In this chapter we will explore the theme of the actuation of responsibility within cybernation. Under this theme we intend to examine the way that responsibility actually takes shape in the cultural situation that develops under the impact of cybernation. The chapter shall be organized around two topics. First, we shall examine responsibility as it is actuated in cybernetic engagement. When man works, i.e., when his power and energy are extended creatively in order to constrain the environment, one dimension of responsibility is seen. When man links his energy dynamics to those of nature, through his sinews or senses or his mind, he works. He is engaged. In this section we shall examine not only the problem of work but also the activities of communication and the arts. The second section will deal with the actuation of responsibility as man is disengaged from the productive relation to the environment. Here the traditional problems of leisure time will be examined.

Man's work relation to environment and his conceptualization of that engagement are shaped by his spiritual perception of reality. The meaning and goal of man's life has been integrally related to his work in the Protestant tradition. The concept of work is related to the correlative concept of time which, as we have noted, has a characteristic urgency in the world history affected by that Christian faith.[3] Man must use his time productively if he is to live purposefully. At this point we understand how deep anthropological problems arise in the development of cybernation. Although man remains in the productive process at the point of decision-making, planning, and intellection in general, he is gradually removed from the visible "result seeing," "finished product fulfillment" phase that work has traditionally had for man. In this exoneration man often undergoes a loss of dignity in self-concept.

The primary source of the intense work-consciousness of Western man has been variously labeled. Some contend that the Western European environment has forced man to work lest he freeze in winter. Historical arguments are more credible. The strongest impetus for the work-energy that has characterized Western man is the concept that he is the co-worker with the "Maker of heaven and earth." Also important is the correlative concept of time that has informed that culture. The Industrial Revolution and the subsequent cybernetic revolution spring directly from an urgent view of the value of time. When technological man today programs an automated factory with computers that process data at incredible speed to hasten production, he responds to a compulsion that inheres in his tradition, to get things done swiftly and accurately.

Our culture has problems welcoming cybernetic disengagement and mass leisure. Ethically speaking, the concept of time has given positive valuation to the virtue of industry and negative valuation to sloth or time-wasting. The Puritan ethic ingrained into man's being the notion that he worked either in order "to consume time," i.e., "Satan finds some mischief for idle hands to do," or as compensation. In the latter case he was either working out the consequences of his sin or he was striving for the justification that hard work bestows. In this understanding it is work alone that justifies rest. The concepts are based on the rhythm of the Sabbath week when God worked, then rested. Concomitantly, only when man labors is he compensated with rest from his labors.

Margaret Mead argues that in the cultural substance

of Western man there "... runs a persistent belief that all leisure must be earned by work and good works and seen in the context of future work and good works."[4] In other words, man works because he must work to pay back some indebtedness and he should work in order to earn some future reward. In each case the motivation is related to the use of time.

In the background of this admittedly distorted view of work lies a view which must be rediscovered, namely, the Christological understanding of time. In Christ, the whole of creation in its spatiality and temporality is being redeemed. Oscar Cullmann in his classic study of the New Testament understanding of time and history has shown the meaning of the fact that the God revealed in Christ is eternal. This means that he is before all things and antecedent to all things. In this setting, time is given a certain valuation. The theological significance of time for Cullmann is contained in the fact that God not only contains human life in pre-existence and predestination, but in him future events are already comprehended in significance.[5] With this cosmic understanding of God and his relation to history and life, time carries a compulsion that man must heed if he is to be responsible to God who stands at the beginning and end of all things.

It is in the nations influenced by Calvinism where the Puritan ethos took hold that this, time-consciousness, and the virtuous valuation of industry, took place. The belief in the sovereignty of God coupled with the predestinarian view of history imbued time with a value that is still very much a part of our technological understanding. One interesting point has been overlooked in the literature of cybernation. The point was hinted at when Einstein in his commentary on the theory of relativity suggested that faith is fundamental to any mathematical projection. Simply stated the idea of feedback rests on an understanding of causality and time which is forged in the crucible of Christian theology. In cybernation we initiate an action through some electronic device. This stimulus facilitates some action, and the reaction which follows engenders some change in output. Although simultaneity is also at the base of the phenomenon (i.e., a spatial rather than temporal analysis of cause and effect), the basic underlying presupposition is that time progresses with meaning and predictability rather than mere chance. A good case can be made for the argument that a Christian understanding of time is fundamental to the concept of feedback in cybernation.

With these background notes on "time and consciousness" from the perspective of theology we can move to a discussion of work as it is seen from the perspective of cybernation. In the midst of man's work he is engaged in the creative task of renewing the earth and calling into the present the future. No matter how menial the task, man has had the feeling that in his work he stands at the cutting edge of the coming world. He is participating in this livelihood not only for the sake of providing for himself the necessities of life but he is participating in the exciting endeavor of creating the new world of tomorrow that is being revealed to man now as he engages today in the creative task.

Bertrand Russell, cynically, but with much justification, asks, "What is work? Work is of two kinds: First, altering the position of matter at or near the earth's surface relatively to other such matter; second, telling other people to do so."[6] The pungency of Russell's wit overshadows the fact that work is far more complex, dealing with a far greater spectrum of occupation than the distance from the ditch-digger to the foreman. Even the proposal to separate men into two groups: laborers and organization men, supposingly exhausting all human enterprise in the catagorization, is too simplistic. Men are all involved in a work situation in which organization and productivity, boredom and creativity, meet in perpetual interface. Although the organizational task certainly predominates in the cybernetic era, who can say that this task is completely devoid of creativity. On the other hand, only the fool would contend that only manual labor is invigorating and "good for the soul." In a complex world where all men are increasingly interdependent on all other men (particularly in the electronic village where events on one side of the world have simultaneous impact on the other side), the organizational task, of course, predominates. Only a world that is highly organized, one in which most men fulfill their vocations not in hand work but in administrative tasks, only this world can face the challenge that the technological age offers. The anthropological question that rises at this point is this: What does this do to man's self-understanding as this is related to his work?

Of course, man must resist "personal functionalization,"[7] as Harvey Cox has noted. The man who becomes the slave of his gadgets or the genies responding to the wish of his apparatus ceases to be a meaningful servant in the cybernetic age. The genius and the opportunity of cybernation is that man, guided by a responsible understanding of the present and of the desired future, can direct that future rather than be directed by it. If man forfeits this initiative he has only himself to blame for the catastrophic results.

The ethical challenge of the twentieth century is to grasp this new-found capacity of control and direct environmental process, moving it in a responsible way. The fact is that our life and the life of our society is being planned and directed. The questions, as Cox notes, are: "By whom is it planned? For whom is it administered? And how well is it being done?"[8] If the social unrest that has marred our decade has anything to say it is surely this: The planning at the societal level *has not been* and *must be* directed to the need-fulfillment of its people. They must be guaranteed the optimum chance of fulfilling their personality and their God-given capacities. This, of course, refers directly to the problem of work. Man needs to see himself involved in the feedback loop where needs are being met and aspirations are being heard and fulfilled. When this is not the case a man becomes estranged from the historical process and "tunes out," to use good cybernetic language. He lives in frustration because his creative capacity is not given a chance to exert itself. The most tragic expression of the alienation of our time is the massive movement to "drop out." The international press reported recently the fact that one-fourth of all Americans use barbituates of some kind.[9] This statistic, compounded with those of alcohol and drug usage, support the thesis that man, shaped by the cybernetic revolution, finds little meaning in his present occupation or preoccupation. He seeks to pull out of the time continuum and live for the intensity of experience that the present can offer.[10]

Work, however, in the cybernated context can be rich and rewarding if man can free himself from the distortions of the Protestant work ethic which asks only, "What did you produce today?" The genuine rigor of that ethic asks, "What did you do today for your fellowman?" It is at this point that the whole question of responsibility in cybernation gathers. Man is released in this new age to exist for self-preservation and self-provision and can rationally expend himself for the good of his neighbor if he so chooses. Granted, the gratification of seeing the response of personal appreciation may be removed in the impersonalism of the cybernetic situation, but perhaps paternalistic gratification was an inferior motivation all the while. Here, perhaps, the organizational tasks can be seen in a renewed fashion. As man freely gives himself to the cybernetic task, throws himself into the highly demanding arena of making decisions under the scrutiny of feedback and the pressure of electronic speed, he has new opportunity to genuinely serve his brother, near and far. Harvey Cox has correctly noted, "From the biblical perspective, the first question is never 'How can I save my own soul, skin, values, or personality?' "[11]

Responsible shaping of values takes into account, first, the truth which man knows (Rom. 1:18); second, the others for whom, through experience, one is responsible (this is extended universally in the age of electronic communications media); and, finally, the consequential perspective on contemplated actions (feedback). In other words, responsible man today must work with a threefold awareness: He has been given the truth of God's revelation, the association with a worldwide family, and the capacity to predict the outcome of certain decisions. In this context, work becomes ethically pregnant because of the way in which responsibility has been intensified.

We have frequently illustrated our concepts with experiences from the space program. Here again this threefold matrix of responsibility can be illumined by an example. The great value questions being raised by the moon program focus on these three dimensions of responsibility. With reference to the knowledge aspect, we now know that we can solve even apparently insurmountable problems if we have the will. We know that we can restore the cities; cleanse the atmosphere, land, and water; even solve the agonizing, paradox-filled human problems of poverty, hunger, and racism *if we decide to.* If we commit the resources, do the evaluation and long-range planning, utilize the technology, and listen to the feedback, we can not only rescue the environment from disaster but again fill the earth, which is the responsible act of subduing the cosmos.

With reference to our responsibility for others, the moon landing has shown us the necessity of international cooperation. Rabinowitch's call for joint Russian-American space flights[12] will probably go unheeded. Yet the hope consensus of men around the world is that space will not be used for exploitation and nationalistic rivalry but rather that it will become a new realm of cooperation, mutuality, and international interdependence.

The predictability of the space ventures of both the U.S.A. and U.S.S.R. has convinced us that we can glimpse and create the future. We can forestall disaster. We can anticipate consequence of actions. We can monitor on Wall Street, in the Kremlin, at the United Nations, the E.K.G. of international response to contemplated actions. We can cybernetically make and keep the peace, if we will.

In the cybernated era work must be liberated from the necessity of visible productivity leading to personal or national aggrandizement to the free awareness that all creativity contributes to the common good. When work ceases to have this value for man, he forfeits his design and control of the process and in his disengagement contributes to the dangerous development of technological super-structures over which he can no longer extend control.

The first step toward such a cultural transformation will come when we understand the rationalization of contemporary industry as a liberating phenomenon as this relates to the way in which we understand our work. Dehumanization is only possible if man polarizes his individual function and fails to see the total picture.[13]

There is a fundamental need on man's part to be involved in the creative process. His identity as a creature has this element as an essential feature. In his createdness there inheres the necessity to be about creativity. In his createdness he stands under God as the subjugator of the earth. When he can pursue this task, he finds fulfillment. When he is frustrated in this desire and he himself takes on a functional identity, he ceases to be a man in the biblical sense of the word. Ceasing to be subject, he becomes the object of manipulation and is dehumanized.

The problem was sensed early in what has proven to be a prophetic book in the light of the technological mania of the Third Reich, written by Hans Lilje. Commenting on an English title, *Principles of Scientific Management,* he saw the danger and the possibility of the polarization and fragmentation of the work process.[14] In a technological time, particularly when destructive capability is of paramount importance to a nation, it is necessary for responsible men to see the whole enterprise. Such comprehensive knowledge is necessary lest the apocryphal story of the Russian factory worker come true. He was told that the factory in which he worked was in the business of building baby carriages. Coming to suspect this, he took a different piece home each night and when he had assembled all the pieces together he found that he had built a machine gun. The ethical challenge in an age when specialization is so necessary is to have sufficient rationalization of industry to permit an overview of the total work process. Perhaps this necessity is somewhat alleviated in the development of cybernation. As opposed to automation more tasks in cybernation are synthetic and analytic in character. The point remains, however, that man can labor meaningfully in the microcosm only when he has an appreciative grasp of the macrocosm. Hegel speaks of the "abstraction" of work as a possibility for the worker who must specialize.[15] Perhaps this is not as important as the placing of his specific productive task in the context of the total enterprise and, indeed, the total common good. The fulfillment of work lies not only in the act but in the purpose or utilization of the act. The man who finds pleasure only in the pure creative act is missing a critical dimension, namely, the dimension where the creative act is shared, received, or enjoyed by another.

Cybernation requires a new life-style on the part of man—a style that can perceive the subtlety and the beauty of intricate electronic processes and glory in them to the same extent that primitive man could revel in his handiwork. In this era the artifacts that prevail in the productive enterprise are those which extend the mental facilities of man. In the understanding and the appreciation that man has of his work he must discover the nobility of that enterprise which extends order into the environment.

Related to the work engagement that man has as he extends facility into the environment cybernetically are the creative communication arts. Here again we meet problems that have ethical dimension, problems that call for an alteration in our mentality. Through the creative arts man exerts the press of his personality to inform the environment.

It is self-evident today that man

> . . . has developed extensions for practically everything he used to do with his body. The evolution of weapons begins with the teeth and fist and ends

> with the atom bomb. Clothes and houses are extensions of man's biological temperature-control mechanisms . . . Power tools, glasses, TV, telephones, and books which carry the voice across both time and space are examples of material extensions . . . All man-made material things can be treated as extensions of what man once did with his body or some specialized part of his body.[16]

Above and beyond this the most sophisticated extension of the rational nature of man is now possible through cybernation. In the age of advanced communications media the former simple relation of man to his environment is changed. A new context of perception develops which carries a new dimension of responsibility. Marshall McLuhan locates the new relationship in terms of a transition from closed systems to open systems. In former times the human facility was extended only in its particularity. It participated only in the unique feedback loop of that particular sense. Now the complex immediacy of electronic communications creates a new situation. "Now in the electronic age, the very instantaneous nature of co-existence among our technological instruments has created a crisis quite new in human history. Our extended facilities and senses now constitute a single field of experience which demands that they be collectively conscious. Our technologies, like our private senses, now demand an interplay and ratio that makes rational co-existence possible."[17]

One can cite many ways in which the contemporary electronic media have served to internalize and intensify experience which contributes to a new context of responsibility. The film has served as a masterful media to internalize the exterior world. Television has rendered distant events immediate to the experience. When through instantaneous feedback one witnesses the assassination of John Kennedy, of Lee Harvey Oswald or Martin Luther King or Robert Kennedy, one feels part of the event. Perhaps this is one of the reasons for the discussions of the collectivity of the guilt of these crimes. As the world gathered around these scenes in the gigantic electronic living room of the world, participating in the full range of action and emotion, is it strange that one so frequently heard the words "we are all responsible" for the death of a King or a Kennedy? Now certainly, this statement is true in the light of the collective social evil which fosters both the political desperation and the violent atmosphere within which the events took place. But certainly the simultaneous feedback of electronic media has also served to call forth this intensified perception of responsibility.

On the other hand, one cannot deny the impact of the electronic media on the urban unrest that breaks out simultaneously in scattered cities around the United States or the student revolt around the world. One feels sympathy and camaraderie in causes and events despite the distance barrier of thousands of miles. Here again we see the necessity of placing responsibility in the context of rationality. The greatest ethical challenge of cybernation lies at the point of value formation. McLuhan recognized this when he wrote:

> . . . modern technology presumes to attempt a total transformation of man and his environment. This calls in turn for an inspection and defense of all human values. And so far as merely human aid goes, the citadel of this defense must be located in the analytical awareness of the nature of the creative process involved in human cognition. For it is in

> this citadel that science and technology have already established themselves in their manipulation of the new media.[18]

The universal participation in the moon landing is an example which bears out McLuhan's thesis. Most of the world was engaged in the experience. The creativity, the adventure, indeed the accomplishment of the feat were widely shared through universal hook-up of communications media. In this case the participation was at the psychic level of joy and hope rather than guilt and shame. That electronic participation is valuative is seen in the contrast of the engagement of a large segment of the American populace in the space event, with proportionate disengagement in the Vietnam conflict; both events mediated to our attention through communications. Man distinguishes ethically through the quality of his participation.

The creative process can be best understood in its ethical fullness as it is seen as a dimension of the divine activity of renewing and humanizing the world and the correlative human endeavor of subduing the cosmos, bringing it under rational control and communication. As the creative arts become media to enrich the experience and as communications media inform the mind and sensitize the conscience, they become vehicles of the divine purpose.

We can speak of responsibility in terms of the one using the media or the one used by the media. Summarily, our desire is for a free ethic. The most dangerous limitation comes when some casuistic principle is introduced. When the answer to what is good in communications is answered by party line, or what will further revolution, or the question "What will harm our youth?" Free creativity alone is responsible. This enterprise elicits the optimum in feedback. It seeks to render reality in its fullness to the experience and bring the full press of the ethical, rational substance of human experience to inform the environment. Only when this openness prevails can the human enterprise of cybernation participate in the divine energy dynamics of the universe.

An interesting contrasting view of responsible creative engagement in the environment is again afforded by the thought of Karl Marx. According to primitive Marxism, man reaches for his essence as he loses himself in the work process for the sake of the universal brotherhood.[19]

Contrasted to this anthropology the Christian world view sees the extension of man in productive creativity as an effort for the other, who is always concrete and never abstract. He participates in the renewal of the world through this endeavor, but this renewal has its direction and impetus in God and not in any intrinsic dynamism. Finally, his creativity is working to the end of the glorification of the Creator God and the elevation of man only insofar as this contributes to the former goal. What Marx calls "the resurrection of nature" is a divine work implemented through human creativity. Man is not perfected in the social-historical process as Marxism contends. Man endures in his ambiguity and is not necessarily redeemed in the process of the technological mastery of environment. With his creativity he can serve or destroy self, society, and world.

Responsibility, in conclusion, is a frank recognition of man's place in the world, his assertion that there is no internal dynamism which renders him helpless, no inevitable progress that obliterates freedom: Responsible man must be involved in the creative enterprise under the full terror of the divine demand, knowing that God has placed him in this world not by accident but for the fulfillment of his creative design for the cosmos.

It is insufficient to discuss responsibility only in the context of engagement. In the contemporary period a great responsibility faces mankind in the new life of disengagement from the productive process that cybernation offers and frequently demands. Perhaps as great a problem as the alienation through work (*entfremdete Arbeit*: Marx), is the alienation that man will increasingly experience as he is disengaged from the productive process. Anyone who has seen a computer replace an office full of employees, or watched a prefabricated building rise while bricklayers stand idle, or witnessed a prematurely retired man slowly die of boredom knows the ethical dimensions of disengagement. The problems of leisure and play need analysis in the light of cybernation.

Leisure is not a simple phenomenon. To understand it in the context of this book we must see it from several sides. To do this we will discuss the phenomenon under the rubrics of leisure as recreation, as escape, as fulfillment, and finally in the theological context of rest. A preliminary necessity is to define leisure. There are two fundamental meanings. First, the word means freedom or opportunity to do something that is desired. When we say that "he was at leisure to go sailing," this initial meaning is conveyed. Second, the word means release from occupation, unoccupied time. This sense conveys the sense of disengagement. In the fullest definition then, leisure is being freed from something to something else. Here we can see the re-creative possibility of leisure time.

To render leisure time creative seems to be one of the greatest challenges of our time. Free time easily degenerates into misuse. Historically speaking, Denney has shown that "Rome gradually generalized the aristocratic play-force of the Greeks into a coliseum mob, and Byzantium made the Hippodrome political."[20] The price America has paid for the sinister pleasure of wandering vagrants is great. One need only recall the life-style of the now-famous assassins or mass-murderers to document this unfortunate fact.[21] To transform the life energy that is released in leisure into creative channels of expression remains a great ethical problem.

In order to approach this problem a new understanding of the relationship between work and leisure is necessary.[22] In Western mentality we have postulated a rhythm to and from work which sees the former as creative and the latter as re-creative. We exhaust ourselves in our work only to be refreshed by the Sunday drive in the country which prepares us for more work during the week. Margaret Mead notes the way that the two are joined attitudinally for the Western man: "The word recreation epitomizes the whole attitude of conditional joy in which the delights of both work and play are tied together in a tight sequence."[23] In the activistic work ethic of the West there often appears an unhealthy cycle that has practical and theoretical consequence in the life-style of the highly industrialized nations. The attitude has a rhythm that moves from hard work, understood as good works, to recreation which gives the energy for more hard work which, of course, stimulates greater good works. The theological critique of this kind of conceptualization is evident. Man does not exist to work. The essence of his existence is the imperative which in the Christian faith becomes the indicative to glorify God in life and to live in compassion for the neighbor. The basis for the work necessity lies in the Fall. Man wrests his bread from the earth in the sweat of his brow since the Fall. Work impulse is not located in redemption. The redeemed man moves in and toward a new creation where toil ceases and his work is done. This is not to say that the 32-hour week is a sure sign of the parousia. The point is that in our intake-exhaust, recreation-creation, rhythmic understanding of the work/leisure dialectic is artificial.

To be disengaged from the productive process is not necessarily to be disengaged from the energy dynamics of life. Man's free time holds the possibility of re-creative growth through interpersonal feedback and communication as well as artistic informing of the environment. It is when man extends his creativity that a feedback loop is created. He cannot be refreshed in a static state. Even his stillness must have a "giving" character if it is to be re-creative. Here we speak of another rhythm pattern. This pattern sees energy release and reception as the possibility of each creative moment. Profound illustrations of this rhythm pattern are afforded by Karl May (1842-1912), a great German storyteller, and "The Birdman of Alcatraz," the great scientist-ornithologist. These men, though cut off from the normal communications network while imprisoned, exhibited unusual creativity in this disengagement. Each of these great artists took the information available (a library in each case), used it in a highly imaginative way, and initiated a dynamic feedback loop, transforming what could have been a completely exhausting experience into one of high intensity and creativity, producing valuable and legendary material. It staggers the imagination to think of the rich experience that inhered in these environs that were so devoid of normal experience.

Even greater, of course, is the possibility of re-creation and refreshment that divine inter-communication affords. How pregnant is that moment of leisure when in stillness one's being stands confronted with the knowledge of God (Ps. 46:10). The concrete meaning of worship and communion is found in that communicative relationship in which God addresses and man understands and praise redounds to glory. Here, in fact, we are up against the most intensely re-creative feedback loop accessible to the human range of communication. Here the being of man is energized by the source of life. Here is the energizing fountain from which he came, in which he lives, and to which he goes. Here the heart of the matter is sensitized.

The cybernetic revolution is a release and an extension of the capacities of communication energy on man's part. The heart of the Christian doctrine of man in creation is the radiation of the communicative energy of God into the cosmos. The contact of divine and human communication energy is an electric moment of creation, filled with power and re-creative energy symbolized by Michelangelo in the Sistine Chapel creation fresco. The moment is re-creative because of the human fulfillment that issues from the divine satisfaction. Emil Brunner locates this when, with reference to man's dominion over nature, he writes, ". . . in man alone can God truly glorify and communicate Himself, because here alone can His love be received by an answering love, because here alone can His Word be answered with a free response."[24]

In the activity of taking dominion over nature or subduing the cosmos (Gen. 1:26), man is re-created as he stands at the point of communicative interplay with God. As man glorifies God, God glorifies man and nature is renewed. As the work of man's hands is established (Ps. 90:17), the cosmic will of God is fulfilled.

This is the re-creation of physical and mental activity where the fullness of the human capacity is exercised. Ultimately, such leisure can afford the microcosm of Sabbath sensitivity where one stands under the Word of God, offers praise, and is renewed by forgiveness and grace. Man must be disengaged so that he can be engaged and re-created at this deepest level of intercourse. The biblical meaning of re-creation is the contemplative withdrawal in the moment of creativity which gives thoughtful purpose to the continuing creativity (Gen. 1). Just as in the pattern of Yahweh's work in creation, the moment of withdrawal is the necessary oasis in the life that would be spiritually sensitized and sustained.

Leisure time can also be misconstrued and used as escape. Although there is certain merit in the new asceticism in which leisure is seen in terms of disciplined creativity,[25] it is very evident today that the problem has taken on negative dimensions. The leisure class analyzed by Veblen has now proliferated so that the activity is no longer exclusive. While in the industrial engagement the Luddite phenomenon (workers so enraged that they smash their machines) has disappeared, so that workers passively accept the self-manipulation that production demands; it is boredom and complacency that often prevail in leisure time.

A very interesting contrast to the traditional leisure problem is emergent today. Historically the workers have labored for the free time of the elite class whose leisure they enjoyed vicariously. Today the worker is released more and more as the organizational elite assumes a greater part of the productive process. Huxley, in *Brave New World,* suggests this theme as the lower classes are given over to blatant hedonism while the ruling "Alphas" alone receive the sophisticated pleasures of thought and work. The theme intensifies in the more recent science fiction of Frederik Pohl, where the status of the upper classes is signified by the fact that they have to do more work, live in smaller houses, and have fewer robots working for them.[26] Leisure in a highly cybernated era can indeed become a curse if it is interpreted and used solely as a means of escape. The answer to the problem does not lie in the "human engineering" employed in some factories where leisure enticements are used to make the workday more productive. These attempts to ameliorate the drudgery by lighting effects and music, by providing social settings and night classes in the vices of the managing elite, indeed all attempts to weave the workers into a fabric which has community, have not been well received.[27]

It appears that a shift in mentality is necessary for both the toiling classes and the leisure masses. With regard to the toiling class there must be a broadening of the base of administration and organizational management so that the worker, his needs, and his aspirations are incorporated into the informational loop that shapes that enterprise that in turn shapes to a great degree his life. This measure would also release the organizational men from the threatening isolation of decision-making. The leisure classes must find new meaning in engagement in their work and also see the positive opportunity that leisure time affords. There is danger if the frantic escape to the suburbs both physically and psychologically continues. The deadening privacy, the disengagement, and the denial of communal responsibility threaten the social fabric of the industrialized nations.

Cybernation offers a new mode of understanding leisure and work whereby both can be given new meaning as work release is effected by computers which share organizational burden. The result is that man's leisure relation to environment is intensified and personalized through greater feedback sensitivity.

The work compulsion that is ingrained in our mentality must diminish, as well as the retreat compulsion which is the contemporary reaction to that former necessity. Instead today we need a new concept of personality which recognizes the rewards that creativity affords, yet perpetually guards against compulsions that regularize that creativity. Escape then is necessary when it is expressed as asceticism. The rational disengagement from high pressured activity, which has lost the measure of contemplation that is necessary for meaningful technological progress, is good. Retreat or escape becomes dangerous when it seeks to "drop out" from the bristling network of energy dynamics, which, whether we like it or not, is the environment of modern man. Responsibility demands engagement. Engagement is responsible only as it is aloof, contemplative—grasping the ends of the process in terms of purpose and meaning.

In his masterful study of nihilism, Helmut Thielicke shows the way in which man, alienated from his identity and purpose, can even approach the sickness unto death in negative escape. When disengagement takes on an incommunicado character or withdrawal character, it becomes schizoid. With an illustration from psychiatric experience he relates an existential progression that can occur in negative disengagement. The sickness is expressed in this way: *"Ich bin nur eine Maschine, ein Automat. Nicht ich bin es, der empfindet, spricht, ißt, nicht ich, der schlaft. Ich existiere gar nicht mehr. Ich bin nicht. Ich bin tot . . .*"[28] When leisure becomes an ingathering, integrating experience it is healing. Energy is released and restored and life is renewed. When, however, the disengagement focuses internally it can become pathological. Escape from the communication network, particularly the interpersonal communications network, fractures and dehumanizes the personality.

In terms of solution we are talking here primarily about a frame of mind, an attitude of engagement or disengagement. The great tragedy of cybernation is the disengaged man, one who, in Ernst Korff's terms, belongs to the "*innerliche Arbeitslosen* [inner feeling of loss]."[29] Here is a poverty that can only issue forth in despair and depersonalization. Here the only way out is the integrity that is the gift of faith. Thielicke goes on in the aforementioned volume to note the positive possibility in the proper feeling of nothingness, where confidence in self-control yields submissively to the gracious control of the divine spirit. Here is the positive possibility of disengagement.[30]

Leisure disengagement is properly conceived of as a channel to fulfillment. If the work engagement offers only a partial realization of the fulfillment of the spectrum of human capacity, surely leisure should serve to awaken these latent capacities.

In the little devotional work of Hans-Rudolf Müller-Schwefe, he notes that free life in the technological era is fulfilled only when the ultimate capacities of communication are activated. "Life on the technological level is only possible when man spiritually and ascetically uses the freedom that is his for the spiritual quest. It is only the spiritual quest that can assure his freedom."[31] It is in asceticism in this meaningful sense rather than in escape that fulfillment is found. A discipline "technological asceticism" is necessary in this age, says Harvey Cox, "that will prevent our becoming captives of our gadgets."[32] What is the characteristic of this disengagement which is fulfilling? First of all, it is escape to rather than escape from. Korff notes that free time for the factory worker holds the capacity of fulfilling the interpersonal and activity anticipations that are denied him in the factory experience.[33] It is only as these free hours are filled with desire fulfillment—family, sports, etc.—that they become hours of creative disengagement. Second, these must be hours of discipline, wherein one energetically seeks to enlarge this or that capacity which lies latent in the one-dimensional character of his work. Finally, he must seek the refreshment of the search for and the service of God. Here alone is the spiritual energy that can renew the exhaustion of man (Ps. 118).

Fulfillment then is the exercise of the various interests, gifts, and capacities of man that are not given channels of expression in the productive process. These energies are not merely expended, they are rewarded, refreshed, and invigorated by those divine powers which reward search with solution, service with satisfaction.

Hans-Eckehard Bahr, in his interesting study on leisure, has shown how the leisure class has not found satisfying liberation in mere freedom, but rather has arrived more certainly at the conclusion that something is missing.[34] He goes on to suggest ways that the church can overcome its suburban captivity by again reaching for meaningful sources of proclamation and service. Here, at the point of understanding what the Sabbath is, lies the most fruitful option for the man who seeks meaningful disengagement from the cybernetic process. A genuine renewed understanding of "Sabbath rest" as interlude *from* and contemplation *of* work is needed in our time.

In the creation understanding of the Sabbath as well as in the sensitive periods in the history of theology, the Sabbath is conceived of as the apex of the rhythm pattern of creative life. It is here where worship restores the directness of the God-man relation, where proclamation renews human life under the Word of God, and where the dynamics of the divine Spirit are acknowledged and received. The Sabbath is not a pull-out from activity. It is not a somber day when, bedecked in black, one trembles before the judgmental God. The Sabbath is rather the intensified point of the life-rhythm pattern where one acknowledges the source, sustenance, and destiny of his life. Here the intense feedback loop of grace and love is acknowledged and submitted to. Here the cycle of love and communion weaves the human family together with God in that life-fulfilling moment when life's purposes are clarified (Ps. 84).

In conclusion we claim that responsibility seeks to enter profoundly into the feedback loop of energy renewal both in engagement of work and in the disengagement of leisure. Actuated responsibility thus takes seriously the fact that one's work is ever concomitant with the One "who works for good with those who love him" (Rom. 8:28). It also knows that leisure is rich in meaning because of the fact that God "rested . . . from all his work" (Gen. 2:2).

We have sought to interpret the phenomenon of cybernation, which is more correctly labeled a mood than a movement. We have sought to interpret theologically the dimensions of this mood. We have sought to examine the context of cybernation as this shapes responsibility.

Much, of course, has been overlooked. The cybernetic impact on tensions of the world has only been suggested. The enhanced facility of the computer today as it can serve to diagnose and give the directions to alleviate these problems is a formidable topic in itself. This capacity has given the language and literature what may be the most important word of our time, "planning." The ethical dimensions of food distribution, peace research, and natural resource exploitation are only a few of the related problems within the scope of cybernetic analysis and solution. These sweeping social-ethical questions require individual attention which is not possible in the scope of this book. We have not thoroughly dealt with the problems of scientific planning and the social aspects of computerization. Much work remains to be done with regard to this new phenomenon that is going to shape the life of the human family so profoundly in the next decades. It is the author's hope that this work will have provided some helpful avenues of thought as we meet this challenge.

Notes

I

1. CBS news coverage of "The Epic Voyage of Apollo 11," July 20, 1969.

2. *Ibid.*, interview with Robert Heinlein, science fiction author.

3. Robert Theobald, "Technology and New Images," *motive*, XXVII, Nos. 6, 7 (March-April 1967), p. 5.

4. Norbert Wiener, *God and Golem, Inc.: A Comment on Certain Points Where Cybernetics Impinges on Religion* (Cambridge, Mass.: The MIT Press, 1964), p. 99.

5. *Ibid.*, p. 1 (Intro.). (In Jewish legend, golem is an embryo Adam, shapeless and not fully created; hence a monster, an automaton.)

6. *Stephanus Thesaurus Graecas Lingua*, Vol. IV, pp. 2053-2056.

7. A. M. Ampère, *Essay on the Philosophy of Science*, 1843, cited in C. R. Dechert, "The Development of Cybernetics," *The American Behavioral Scientist*, VIII (June 1965), p. 15.

8. J. C. Maxwell, *Proceedings of the Royal Society* (London: Royal Society, 1868), Vol. XVI, pp. 278-283.

9. Norbert Wiener, *Cybernetics or Control and Communication in the Animal and the Machine* (Cambridge, Mass.: The MIT Press, 1948), p. i.

10. Donald N. Michael, *Cybernation: The Silent Conquest* (a report to the Center for the Study of Democratic Institutions) (Santa Barbara, Calif., 1962), p. 47.

11. "Biological Aspects of Cybernetics" (Moscow, 1962), *JPRS*, 19, 637, p. 17, cited in Dechert, *op. cit.*, p. 16.

12. Wiener, *God and Golem, Inc.*, *op. cit.*, p. 3.

13. *Ibid.*, pp. 74-75. The development of artificial hands in the cybernetics centers of Russia and the development of iron lungs are examples. "Let us suppose that a man has lost a hand at the wrist. He has lost a few muscles that serve chiefly to spread the fingers and to bring them together again, but the greater part of the muscles that normally move the hand and the fingers are still intact in the stump of the forearm. When they are contracted, they move no hand and fingers, but they do produce certain electrical effects known as action potentials. These can be picked up by appropriate electric motors, which derive their power through appropriate electric batteries or accumulators, but the signals controlling them are sent through transistor circuits. The central nervous part of the control apparatus is generally almost intact and should be used. Such artificial hands have already been made in Russia . . ."

14. Daniel S. Halacy, *CYBORG: Evolution of the Superman* (New York: Harper and Row, 1965), p. 205; Harold Hatt, *Cybernetics and the Image of Man* (Nashville: Abingdon Press, 1968), p. 304.

15. John Von Neumann, "Machines and Man," cited by the editor, *Scientific American* (New York: Simon and Schuster, 1955), p. 146.

16. Arthur C. Clarke, *2001: A Space Odyssey* (New York: Signet Books, 1968).

17. Von Neumann, *op. cit.,* p. 147.

18. Neville Moray, *Cybernetics* (New York: Hawthorn Books, 1963), p. 116.

19. Norbert Wiener, "Some Moral and Technical Consequences of Automation," *Science,* CXXXI, No. 3410 (May 6, 1960), p. 1355.

20. Erik Barnauw, "McLuhanism Reconsidered," *Saturday Review,* XLIX, No. 30 (July 23, 1966), p. 19.

21. Marshall McLuhan, *Understanding Media: The Extensions of Man* (New York: McGraw-Hill Co., 1964), p. 21.

22. William R. Cozart, "Cybernetics: Meta-Image of the Twentieth Century," *Ecumenical Institute Newsletter,* III, No. 2 (August 1966), p. 2.

23. Hans Jonas, *The Phenomenon of Life: Toward a Philosophical Biology* (New York: Harper and Row, 1966), p. 110.

24. Max Weber, *Gesammelte Aufsätze zur Religionssoziologie* (Tübingen: J. C. B. Mohr, 1920-1921), 3 vols.

25. John Calvin, *Institutes of the Christian Religion,* Vol. 1, ed. John T. McNeill (Philadelphia: The Westminster Press, 1960), pp. 52-53, italics mine. (For interpretation of this point of God contemplated in his works, see L. Wencelius, *L'Esthetique de Calvin* [Paris: Société des Belles Lettres, 1937], chs. 1, 2.)

26. Harvey Cox, *The Secular City* (New York: The Macmillan Company, 1965), p. 23.

27. John Wren-Lewis, *The Meaning of Technology in a Non-Technology Culture* (unpublished document), World Council of Churches and Rapid Social Change Conference, Odensa, Denmark, August 1958, p. 1.

28. Helmut Thielicke, *The Freedom of the Christian Man* (New York: Harper and Row, 1963), p. 125.

29. Nickolai Berdyaev, *The Meaning of History,* cited by John Baillie, *Natural Science and the Spiritual Life* (New York: Charles Scribner's Sons, 1952), p. 30.

30. McLuhan, *op. cit., passim.*

31. Arend Th. Van Leeuwen, *Christianity in World History* (New York: Charles Scribner's Sons, 1966), pp. 158 ff.

32. Richard Shaull, "Revolutionary Change in Theological Perspective," *Christian Social Ethics in a Changing World,* ed. John C. Bennett (New York: Association Press, 1966), p. 23. (Study documents prepared for The Church and Society World Conference of the World Council of Churches held in July 1966 in Geneva, Switzerland, on the theme "Christians in the Technical and Social Revolutions of Our Time.")

33. Jacques Ellul, *The Technological Society,* tr. John Wilkinson (New York: Alfred A. Knopf, Inc., 1964), p. 306.

34. Robert Theobald, "The Cybernated Era," *Vital Speeches,* XXX (June 30, 1964), p. 636.

35. Friedrich Gogarten, *Verhängnis und Hoffnung der Neuezeit* (Stuttgart: Friedrich Vorwerk Verlag, 1953), p. 229.

36. Paul Tillich, *The Protestant Era* (Chicago: University of Chicago Press, 1948), p. 220. Tillich criticizes irresponsible projections of another age: "The secular world does not want to return into heteronomy and ecclesiastical servitude. Protestantism stands above this alternative. It has no ecclesiastical aspirations but subjects them, wherever they appear, to the same criticism to which it subjects arrogant secularism, scientific, political or moral. It tries to create a Protestant secularism, a culture related to a Gestalt of grace as its spiritual center."

37. See Bertrand Russell's essay "Let's Stay off the Moon," *Wall Street Journal* (July 16, 1969), p. 6.

38. Helmut Thielicke, *Theologische Ethik I* (Tübingen: J. C. B. Mohr [Paul Siebeck], 1958), pp. 50 ff. (This translation is from the galley proofs of G. Bromiley; the English edition appeared in 1967, Fortress Press, publishers.)

39. E. A. Burtt, *The Metaphysical Foundations of Modern Science* (New York: Doubleday and Co., 1954), pp. 238-239. Burtt expresses the transition in rather harsh words, saying that after Newton emerged

> ... a view of the cosmos which saw in man a puny irrelevant, spectator ... of the vast mathematical systems whose regular motions according to mechanical principles constituted the world of nature ... The gloriously romantic universe of Dante and Milton, that sets no bounds to the imagination of man as it played over space and time, had now been swept away ... The world that people had thought themselves living in—a world rich with colour and sound, redolent with fragrance, filled with gladness, love and beauty, speaking everywhere of purposive harmony and creative ideals—was crowded now into the minute corners in the brains of scattered organic beings ... The world of qualities as immediately perceived by man became just a curious and quite minor effect of that infinite machine beyond.

40. Jacques Maritain, *The Dream of Descartes* (New York: The Philosophical Library, 1944), p. 21.

41. John Herman Randall, Jr., *The Making of the Modern Mind,* rev. ed. (New York: Houghton Mifflin Co., 1940), p. 249.

42. *Ibid.,* pp. 246-247.

43. Benedict Spinoza, *Ethics,* cited in *ibid.,* p. 247.

44. Floyd W. Matson, *The Broken Image: Man, Science, and Society* (New York: Doubleday and Co., 1964), p. 11.

45. Herbert Butterfield, *The Origins of Modern Science* (New York: The Macmillan Co., 1957), p. 120.

46. A. d'Alembert, *Elements de Philosophie,* quoted in Ernst Cassirer, *The Philosophy of the Enlightenment* (Boston: Beacon Press, 1955), pp. 46-47.

47. Ernst Benz, *Evolution and the Christian Hope* (New York: Doubleday and Co., 1966), p. 68.

48. Thomas H. Huxley, "On the Physical Basis of Life," in *Method: and Results: Essays* (London, 1893), pp. 130-165.

49. Jonas, *op. cit.*

50. C. F. Von Weizsäcker, *The Relevance of Science* (New York: Harper and Row, 1964), p. 192.

51. Ludwig Von Bertalanffy, *Problems of Life* (New York: Harper Torchbooks, 1960), p. 202.

52. Norbert Wiener, *The Human Use of Human Beings: Cybernetics and Society* (Boston: Houghton-Mifflin Co., 1954), p. 188. According to Wiener:

> Albert Einstein's remark . . . is of the greatest significance. *"Der Herr Gott ist raffiniert, aber boshaft ist Er nicht."* "God may be subtle, but he isn't plain mean."
>
> Far from being a cliché, this is a very profound statement concerning the problems of the scientist. To discover the secrets of nature requires a powerful and elaborate technique, but at least we can expect one thing—that as far as inanimate nature goes, any step forward that we may take will not be countered by a change of policy by nature for the deliberate purpose of confusing and frustrating us. There may indeed be certain limitations to this statement as far as living nature is concerned, for the manifestations of hysteria are often made in view of an audience, and with the intention, which is frequently unconscious, of bamboozling that audience. On the other hand, just as we seem to have conquered a germ disease, the germ may mutate and show traits which at least appear to have been developed with the deliberate intention of sending us back to the point where we have started.
>
> These infractuousites of nature, no matter how much they may annoy the practitioner of the life sciences, are fortunately not among the difficulties to be contemplated by the physicist. Nature plays fair and if, after climbing one range of mountains, the physicist sees another on the horizon before him, it has not been deliberately put there to frustrate the effort he has already made.

53. Moray, *op. cit.,* pp. 123-124:

> The picture of the capabilities of artifacts, machines which learn, evolve, adapt, and reproduce themselves, may seem at first sight to be a frightening one. And it is one with which we will have to learn to live. But at the same time, we have seen that out of the analysis which cybernetics gives the world, there emerges a new apologia for the uniqueness of man. The way of thinking is unfamiliar as yet, and of the more traditional Christian arguments, there is hardly a trace. But this has a certain advantage. For those dedicated to the use of science not for the manufacture of methods of destruction, but for the turning of the world into a place where there will be a more "human use of human beings," it provides a common picture which . . . shows man to be unique . . . man stands unique—a unique system in a universe of systems . . . the ruler alike of animal and artifact . . . A little lower than the angels crowned with glory and honor.

54. Cited in John A.-Day, *Science, Change and the Christian* (Nashville: Abingdon Press, 1965), p. 70.

55. See Cox, *op. cit.,* pp. 40 ff.

56. *Ibid.,* pp. 41-43.

57. Thielicke, *Theologische Ethik I, op. cit.,* p. 59, my translation. Helmut Thielicke locates this uniqueness christologically! In der Maske meines nächsten geht der Heiland über die Erde. Das ist die "fremde Würde" die ihn

auszeichnet. Jesus Christus ist für ihn gestorben. (Disguised as my neighbor, the Savior moves over the earth. This is the alien dignity: Jesus Christ died for him) Romans 14:15; I Corinthians 8:11.

58. Day, *op. cit.*, p. 11.

II

1. William Shakespeare, *The Tempest* (V.i.181-184) in *The Complete Works of William Shakespeare*, ed. W. J. Craig (London: Oxford University Press, 1964), p. 21.

2. Paul Lehmann, "The Dynamics of Reformation Ethics," *Princeton Seminary Bulletin*, XLIII (Spring 1950), pp. 17-22.

3. Paul Lehmann, *Ideology and Incarnation* (Geneva: John Knox House, 1962), p. 24.

4. Thielicke, *Theologische Ethik I, op. cit.*, p. 699.

5. Calvin, *op. cit.*, p. 840.

6. Cited in Barbara Ward, *Faith and Freedom* (New York: Doubleday and Co., 1958), p. 109.

7. Dietrich Bonhoeffer, *Prisoner for God: Letters and Papers from Prison* (New York: The Macmillan Co., 1960), p. 178.

8. Cozart, *op. cit.*, p. 4.

9. *Ibid.*, p. 6.

10. Pierre Teilhard de Chardin, *The Phenomenon of Man* (New York: Harper Torchbooks, 1959), p. 180.

11. *Ibid.*, p. 243.

12. *Ibid.*, p. 283.

13. *Ibid.*, p. 304.

14. Bertrand Russell, *A Free Man's Worship: Mysticism and Logic*, cited in Burtt, *op. cit.*, p. 23.

15. Emmanuel G. Mesthene, "Religious Values in the Age of Technology" (unpublished manuscript of speech delivered at the World Council of Churches Conference on Church and Society in Geneva, Switzerland, July 1966. The author supplied the original manuscript which was abstracted and printed in *The Saturday Review* [November 19, 1966] and *Theology Today* [January 1967]).

16. *Of Men and Machines*, ed. Arthur O. Lewis, Jr. (New York: E. P. Dutton & Co., 1963), p. xiv.

17. Walter Reuther (from a statement made before the subcommittee on economic stabilization of the Joint Committee on Automation and Energy Resources of the Joint Economic Committee, U. S. Congress), *New Views on Automation*, 86th Congress, second session USGPO, 1960, p. 513.

18. Rudolf Allers, "Technology and the Human Person," in *Technology and Christian Culture*, ed. Robert Paul Mohan (Washington: The Catholic University of America Press, 1960), p. 4.

19. Mesthene, *op. cit.*, p. 5.

20. Henry Clark, "Human Values and Advancing Technology," in *Human Values on the Spaceship Earth* (two papers prepared for a special project in Human Values in a Society of Advancing Technology of the Commission on the Church and Economic Life Division of Christian Life and Mission of the National Council of Churches of Christ in the U.S.A.), 1966, p. 40.

21. Harvey Cox, "The Possibilities of the Christian in a World of Technology," in *Economic Growth in World Perspective,* ed. Denys Munby (New York: Association Press, 1966), ch. 8, quoted in Charles C. West, "Theological Table Talk: The Experts and the Revolutionaries," in *Theology Today,* XXIII (October 1966), p. 412.

22. John P. Merrill, "Letters and Comments: Moral Problems of Artificial Transplanted Organs," *The Annals of Internal Medicine,* LXI (August 1964), pp. 356 ff.

23. Lee B. Lustad, M.D., quoted in "The Technological Revolution," *Progressive Architecture,* XLVI (December 1966), p. 30.

24. *Ibid.*

25. See the journal *Astronautics,* especially the 1967-1968 volumes, for various papers relating to man's adapting to the space environment.

26. *Time Magazine* (June 14, 1968), p. 59. Reports on the robots being manufactured in Connecticut. The typical robot is rigged with a hydraulic arm which can be adapted to the many manual functions now required in industry. The company expects to sell or rent 5,000 of them before 1970. Examples of the work performed are brick-handling in Georgia, painting cars in Detroit, and picking up glassware in Corning, New York.

Norman Schafler, Unimation Company chairman, notes that "the tireless Robots take no lunches or coffee-breaks, and do not care about working more than one shift."

Of course, the machines have not been well received by Labor. A United Auto Workers' leader says that the machines can conceivably be set up "to shake the hands of the men it replaces."

More realistic is the opinion that the machines can be an assistant in the productive process in a way that man is released, not from the job but from the drudgery of task that has worn so many men down through the industrial ages. This view is spoken by a Caterpillar driver: "The work is hot and repetitive . . . for the worker it was just not desirable. For the Robot it's just a job."

27. Michael, *op. cit.,* pp. 11-12.

28. Wiener, *God and Golem, Inc., op. cit.,* pp. 11-12.

29. Dietrich von Oppen, "The Era of the Personal," in *Man and Community* (New York: Association Press, 1966), ch. 10, quoted in West, *op. cit.,* p. 421.

30. Henri Queffelec, *Technology and Religion* (New York: Hawthorn Books, 1964), p. 63.

31. Quoted in *ibid.,* p. 102.

32. Karl Jaspers, "The Individual and Mass Society," in *Religion and Culture: Essays in Honor of Paul Tillich,* ed. Walter Leibrach (London: SCM Press, 1959), p. 39.

33. "John Henry" is an American folk ballad written about 1870. The author is unknown. The ballad can be found in *Of Men and Machines, op. cit.,* p. 194.

34. Will Herberg, *The Writings of Martin Buber* (New York: Meridian Books, 1956), p. 14.

35. C. A. Coulson, *Science, Technology and the Christian* (Nashville: Abingdon Press, 1960), p. 59.

36. Albert Einstein, *Ideas and Opinions* (New York: Crown Publishers, Inc., 1954), pp. 48-49.

37. Coulson, *op. cit.,* p. 104.
38. Mesthene, *op. cit.,* p. 18.
39. *Ibid.,* p. 16.
40. Thielicke, *Theologische Ethik I, op. cit.*, see ch. 1. English edition by William H. Lazareth (Philadelphia: Fortress Press, 1966), p. 697.
41. Thielicke, *The Freedom of the Christian Man, op. cit.,* p. 129.
42. David Cairns, "Brunner's Conception of Man as Responsive, Responsible Being," in *The Theology of Emil Brunner,* ed. Charles W. Kegley (New York: The Macmillan Co., 1962), p. 78.
43. Dietrich von Hildebrand, "Technology and Its Dangers," in *Technology and Christian Culture, op. cit.,* p. 36.
44. John Wren-Lewis, from *Faith, Fact and Fantasy* by C. F. D. Moule, P. R. Baelz, D. A. Pond, and John Wren-Lewis (Philadelphia: The Westminster Press). Quoted in Myron B. Bloy, "The Christian Function in a Technological Culture," *The Christian Century,* LXXXIII (February 23, 1966), p. 233.
45. Henry Clark, "Cybernation: A Challenge to Christian Ingenuity," *Christianity in Crisis,* V (1965-1966), p. 312.

III

1. Paul Tillich, *Dynamics of Faith* (New York: Harper and Bros., 1957), p. 12.
2. For a discussion of this point in literature, see the fine study of Sallie te Selle, *Literature and the Christian Life* (New Haven: Yale University Press, 1966), pp. 65 ff.
3. C. P. Snow, "The Moon Landing," *Look* (August 26, 1969), p. 72.
4. Wiener, *The Human Use of Human Beings, op. cit.,* p. 18.
5. *Ibid.,* p. 162.
6. *Ibid.,* p. 97.
7. *Ibid.,* pp. 51, 52.
8. Eugene Rosenstock-Hussey, *Out of Revolution* (New York: Four Wells, 1964), p. 47.
9. *Ibid.,* p. 50.
10. Andrei D. Sakharov, *Progress, Coexistence and Intellectual Freedom* (New York: Wm. Norton & Co., Inc., 1968), pp. 60-61.
11. Gyorgy Kepes, "Where Is Science Taking Us?" *The Saturday Review,* XLIX (March 5, 1966), p. 66.
12. Ernest Jones, *The Life and Work of Sigmund Freud,* Vol. III: *The Last Phase* (New York: Basic Books, Inc., 1957), p. 441.
13. Norbert Wiener, "A Scientist Rebels," *Atlantic Monthly,* CLXXIX (January 1947), p. 46.
14. Robert L. Sinsheimer, "The End of the Beginning," *Bulletin of the California Institute of Technology,* LXXVI, 1 (March 1967), pp. 1, 8.
15. J. Robert Oppenheimer, "Physics in the Contemporary World," *The Technology Review,* L (February 1948), pp. 202-203.
16. Ian G. Barbour, *Christianity and the Scientist* (New York: Association Press, 1960), p. 84.
17. William Pollard, "The Christian and the Atomic Crisis," *Christianity Today,* III (October 13, 1958), p. 122.

18. Max Lerner, "Technological Change and Social Progress," *Monmouth College Bulletin* (1965), p. 3.

19. Sinsheimer, *op. cit.*, p. 8.

20. See Thielicke, *Theologische Ethik I, op. cit.*, p. 120.

21. James Olds at the University of Michigan reports that in his laboratory the rat will forsake food and water and even a female rat in heat when he has a triggering control that provides electrical excitement to his brain. For an ethical discussion of this, see Elisabeth D. Dodds, *Voices of Protest and Hope* (New York: Friendship Press, 1965), pp. 46 ff.

22. Hudson Hoagland, "Some Biological Considerations of Ethics," in *Technology and Culture in Perspective*, an occasional paper for the Church Society for College Work (Cambridge, Mass., 1967), p. 18.

23. Robert Sperry, "Mind, Brain, and Humanist Values," *The Bulletin of Atomic Scientists*, XXII (September 1966), p. 5.

24. Eugene Burdick and Harvey Wheeler, *Fail Safe* (New York: McGraw Hill Co., 1962).

25. Clark, "Human Values and Advancing Technology," *op. cit.*, p. 46.

26. Hans-Eckehard Bahr, *Verkündigung als Information* (Hamburg: Furch-Verlag, 1968), p. 139.

27. Herbert Marcuse, *One Dimensional Man* (Boston: Beacon Press, 1966), p. xvi.

28. John C. H. Wu, "Technology and Christian Culture: An Oriental View," in *Technology and Christian Culture, op. cit.*, p. 101.

29. Lin Yutang, *The Wisdom of Laotse* (New York: The Modern Library, 1948), p. 310.

30. Wiener, "Some Moral and Social Consequences of Automation," *op. cit.*, p. 1355.

31. Jaspers, *op. cit.*, pp. 40-41.

32. Erich Fromm, *The Heart of Man* (New York: Harper and Row, 1964), p. 57.

33. *Ibid.*, p. 58.

34. Paul Tillich, quoted in Matson, *op. cit.*, p. 18.

35. Thielicke, *The Freedom of the Christian Man, op. cit.*, p. 138.

36. Père Dubarle, quoted in Wiener, *The Human Use of Human Beings, op. cit.*, p. 180.

This article is referred to at length because of the way in which it locates the ethical dimensions of cybernation in the total social context. The problems cannot be meaningfully analyzed in a narrow anthropological or even technological context. Regardless of the way in which the positivists seek to restrict the context of understanding there are issues here that cannot, by definition, be discussed in the context of a highly specialized and compartmentalized view of social development. The passage of Dubarle is quoted in its entirety here for this reason.

> As far as one can judge, only two conditions here can guarantee stabilization in the mathematical sense of the term. These are, on the one hand, a sufficient ignorance on the part of the mass of the players exploited by a skilled player, who moreover may plan a method of paralyzing the consciousness of the masses; or on the other, sufficient good-will to allow one, for the sake of the stability of the game, to refer

his decisions to one or a few players of the game who have arbitrary privileges. This is a hard lesson of cold mathematics, but it throws a certain light on the adventure of our century: hesitation between an indefinite turbulence of human affairs and the rise of a prodigious Leviathan. In comparison with this, Hobbes' *Leviathan* was nothing but a pleasant joke. We are running the risk nowadays of a great World State, where deliberate and conscious primitive injustice may be the only possible condition for the statistical happiness of the masses: a world worse than hell for every clear mind.

In the light of the turbulent unrest that grips our world, reflected clearly in the situation in France in the summer of 1968, which, of course, is merely a symptom of a groping that penetrates all the highly industrialized nations, we must move to discover the meaning of our cultural estrangement from decision-making.

37. Ronald Gregor Smith, *The New Man* (New York: Harper & Row, 1956).

38. Helmut Thielicke, *Man in God's World* (New York: Harper & Row, 1963), p. 132.

39. John Alexander Mackay, "Creative Communion with God," *Presbyterian Life,* XX, 9 (May 1, 1967), p. 13.

IV

1. Hans Schmidt, "Erwägugen zum Vorgang der Projektwissenschaften," tr. Thomas Wieser, Hamburg unpublished document, 1966.

2. Gerhard Von Rad, *Old Testament Theology* (New York: Harper & Bros., 1962), Vol. 1, p. 137.

3. Karl Barth, *Church Dogmatics* (Edinburgh: Black & Sons, 1958), Vol. III, pt. 1, p. 110.

4. Von Weizsäcker, *op. cit.,* p. 50.

5. Ian G. Barbour, *Issues in Science and Religion* (Englewood Cliffs, N. J.: Prentice-Hall, Inc., 1966), p. 42.

6. Von Rad, *op. cit.,* p. 145.

7. *Ibid.,* p. 147.

8. *Ibid.,* p. 138.

9. *The Interpreter's Bible,* ed. George Buttrick (Nashville: Abingdon Press, 1954), Vol. 9, *Acts; Romans,* pp. 520-521.

10. Theodosius Dobzhansky, *The Biology of Ultimate Concern* (New York: The New American Library, 1967), p. 8.

11. J. D. Chenu, "The Need for a Theology of the World," in *The Great Ideas Today,* ed. Robert Hutchins and Mortimer Adler (Chicago: Encyclopaedia Britannica, Inc., Wm. Benton, 1967), p. 61.

12. *Ibid.,* p. 69.

13. Irenaeus, *Adversus Haereses,* in *The Ante-Nicene Fathers,* Vol. I (Grand Rapids, Mich.: Wm. B. Eerdmans Co.), quoted in R. A. Norris, *God and World in Early Christian Theology* (New York: Seabury Press, 1965), p. 90.

14. Tertullian, *Praescripta,* quoted in Norris, *op. cit.,* p. 105.

15. *Ibid.,* p. 109.

16. Tertullian, *Against Marcion,* in *ibid.,* p. 112.

17. John Calvin, *Commentary on the Book of Psalms* (Grand Rapids, Mich.: Wm. B. Eerdmans Publishing Co., 1949), p. 106.
18. *Ibid.*, pp. 105-106.
19. T. F. Torrance, *Calvin's Doctrine of Man* (London: Lutterworth Press, 1949), p. 25.
20. Calvin, *Commentary on the Book of Psalms, op. cit.*, p. 106.
21. *Ibid.*, p. 107.
22. Martin Luther, quoted in Hans Preusz, *Martin Luther, der Kunstler* (1931), pp. 229 ff.; quoted also in Heinrich Bornkamm, *Luther's World of Thought* (St. Louis: Concordia Publishing House, 1958), p. 179.
23. Martin Luther, *Commentary on Galatians,* quoted in Bornkamm, *op. cit.*, p. 180.
24. Martin Luther, *Vom Abendmahl Christi* (1528), quoted in Bornkamm, *op. cit.*, p. 180.
25. Günther Bornkamm, *Luthers Lehre von den Zwei Reichen im Zusammenhang seiner Theologie* (Stuttgart: Gutersloher Verlagshaus Gerd Mohn, 1964), p. 201.
26. Von Martin Seils, *Der Gedanke vom Zusammenwirken Gottes und des Menschen in Luthers Theologie* (Stuttgart: Gutersholer Verlagshaus Gerd Mohn, 1964), p. 201.
27. *Ibid.*, p. 184.
28. Martin Luther, *Dictata,* III, 369.7, quoted in "An Essay on the Development of Luther's Thought on Justice, Law and Society," *Harvard Theological Studies,* XIX (Cambridge: Harvard University Press, 1959), p. 18.
29. *Ibid.*, p. 171.
30. Seils, *op. cit.*, p. 184.
31. Ludwig Feuerbach, *The Essence of Christianity* (Boston: Houghton, Mifflin, and Co., 1881), p. 126.
32. Hugh Ross Mackintosh, *Types of Modern Theology* (New York: Charles Scribner's Sons, 1939), p. 126.
33. Quoted in *ibid.*, p. 82.
34. Karl Barth, *Die Kirchliche Dogmatik,* III/2 (Zürich: Evangelischer Verlag, A. G. Zollikon, 1959), p. 1.
35. *Ibid.*, p. 11.
36. *Ibid.*, p. 19.
37. Paul Tillich, *Systematic Theology,* III (Chicago: University of Chicago Press, 1963), p. 59.
38. *Ibid.*, p. 73.
39. *Ibid.*
40. Robert Jungk, *Brighter Than a Thousand Suns* (New York: Grove Press, Inc., 1958), pp. 56-57.
41. Tillich, *Systematic Theology, op. cit.*, p. 74.
42. *Ibid.*
43. See William H. Whyte, Jr., *The Organization Man* (New York: Simon & Schuster, 1956), p. 429, and Herbert Marcuse, *One Dimensional Man,* from *Studies in the Ideology of Advanced Industrial Society* (Boston: Beacon Press, 1964), p. 260.
44. Tillich, *Systematic Theology, op. cit.*, p. 74.
45. Dietrich Bonhoeffer, *Ethics* (New York: The Macmillan Co., 1955), p.

74. See also German edition, *Ethik* (München: Chr. Kaiser Verlag, 1949), pp. 71 ff.

46. *Ibid.,* pp. 74-75.

47. Wolf-Dieter Marsch, "Kybernetik und Ethos," in *Sonderheft der Pastoraltheologie* (56 Jahrgang, Heft 4, April 1967), p. 179 (my translation).

48. Chardin, *op. cit.,* p. 313.

49. Jürgen Moltmann, *The Theology of Hope* (New York: Harper and Row, 1967), p. 21.

50. Helmut Thielicke, *How the World Began* (Philadelphia: Fortress Press, 1961), p. 67. German edition *Wie die Welt Begann* (Stuttgart: Omell-Verlag, 1960), p. 75.

51. Clark, "Human Values and Advancing Technology," *op. cit.,* pp. 53 ff.

52. Schmidt, *op. cit.,* p. 19.

V

1. Ralph Waldo Emerson, "Works and Days," quoted in *Of Men and Machines, op. cit.,* p. 75.

2. Karl Marx, "Thesis on Feuerbach," quoted in Bernard Delfgaauw, *The Young Marx* (New York: Sheed & Ward, 1967), p. 50.

3. Freeman Dyson, "Human Consequences of the Exploration of Space," *Bulletin of Atomic Scientists,* Vol. XXV, No. 7 (September 1969), p. 13.

4. Moltmann, *op. cit.,* pp. 22-23. See also "Ernst Bloch: Messianismus und Marxismus. Einführende Bemerkungen zum 'Prinzip Hoffnung,' " in *Kirche in der Zeit,* 1960, pp. 291-295, and "Die Menschenrechte und der Marxismus." Einführende Bemerkungen und Kritische Reflexionen zu E. Blochs "Naturrecht und Menschliche Wurde," in *Kirche in der Zeit,* 1962, pp. 122-126.

5. Van Leeuwen, *op. cit.,* p. 124.

6. Benz, *op. cit.,* p. 124.

7. Hans Lilje, *Das Technische Zeitalter* (Hamburg: Furche-Verlag, 1928), p. 76; Paul Tillich, "Logos und Mythos der Technik," an address delivered on the occasion of the celebration of the 99th anniversary on the establishment of the Technische Hochschule, Dresden. Published in *Logos,* XVI (1927), p. 364. Quoted in *ibid,* p. 252.

8. Benz, *op. cit.,* p. 127.

9. Augustine, *The City of God* (New York: Doubleday & Co., Inc., 1958), p. 524.

10. *Ibid.,* p. 528.

11. *Ibid.,* p. 529.

12. *Ibid.,* p. 530.

13. *Ibid.*

14. Benz, *op. cit.,* p. 128.

15. Van Leeuwen, *op. cit.,* p. 408.

16. McLuhan, *op. cit.*

17. Buckminster Fuller, a visionary architect in America, inventor of the geodesic dome, who contends that environment is susceptible to planning. See especially *Nine Steps to the Moon* (Carbondale, Ill.: Southern Illinois University Press, 1959), and *No More Second Hand God* (Carbondale, Ill.: Southern Illinois

University Press, 1964), where he explores this thesis.

18. See Ernst Bloch, *Das Prinzip Hoffnung* (Frankfurt: Suhrkamp Verlag, 1959), 5 vols., 1657 pp.; see also the commentary by U. Duchrow, E. W. Hirsch, and H. Timm, "Eschatologie der Alternativen," in *Kybernetik und Theologie: Pastoraltheologie, Wissenschaft und Praxis,* 56 Jahrgang, Heft 4 (April 1967), pp. 190 ff.

19. Gerhard Ebeling, *Gott und Wort* (Tübingen: Mohr Verlag, 1966), pp. 43 ff.

20. Hans-Dieter Bastian, "Anfangsprobleme im Gespräch zwischen Kybernetik und Theologie," in *Theologia Practica,* III Jahrgang, Heft I (January 1968), p. 41.

21. See especially Karl Marx, "The Holy Family," and "Ökonomisch-philosophische Manuscripte," in *Frühe Schriften I* (Stuttgart: Cotta-Verlag, 1962), pp. 667 ff.

22. Karl Marx, quoted in Jack Lindsay, *Marxism and Contemporary Science* (London: Dennis Dobson Ltd., 1949), p. 20.

23. *Ibid.*

24. Arnold Toynbee, *A Study of History,* Vol. II (London: Oxford University Press, 1949), p. 178.

25. Lindsay, *op. cit.,* p. 34.

26. Friedrich Engels, "The Role of Labor in the Ape's Evolution into Man," in *Dialectics,* quoted in Lindsay, p. 35.

27. See, for example, Hans Boeck, *Zur Marxistischen Ethik und Sozialistischen Moral* (Berlin: Akademie-Verlag, 1958), p. 39.

28. Helmut Thielicke, "Theologische Befragung des Marxistischen Humanismus," a privately distributed paper. Part of the paper appeared in the volume *Theologie der Anfechtung* (Tübingen: J. C. B. Mohr, 1949), my translation.

29. See the analysis of the concepts *Futurum* and *Adventus* in Jürgen Moltmann's "Antwort auf die Kritik der Theologie der Hoffnung," in *Diskussion über die Theologie der Hoffnung,* ed. Wolf-Dieter Marsch (München: Chr. Kaiser Verlag, 1967), pp. 212-213. See also Carl E. Braaten, *The Future God* (New York: Harper and Row, 1969), pp. 29 ff.

30. Joan Robinson, *Essays on Marxian Economics,* quoted in Lindsay, *op. cit.,* p. 196.

31. G. W. F. Hegel, *Selections* (Berlin: J. Loewenberg, 1929), quoted in Lindsay, *op. cit.,* p. 211.

32. Lindsay, *op. cit.,* p. 212.

33. Seymour Farber and Roger Wilson, *Control of the Mind* (New York: McGraw Hill, Inc., 1961), pp. xii-xiii.

34. Kurt Strunz, *Integrale Anthropologie und Kybernetik* (Heidelberg: Quelle & Meyer Verlag, 1965), pp. 95 ff.

35. Helmut Thielicke, *Theological Ethics: Foundations* (Philadelphia: Fortress Press, 1966), Vol. I, pp. 298 ff.

36. *Ibid.,* p. 317.

37. Immanuel Kant, *Critique of Pure Reason* (London: J. B. Dent and Sons, 1934), p. 457.

38. Moltmann, *Theology of Hope,* pp. 87 ff.

39. Ernst Bloch, *Philosophische Grundfragen zur Ontologie des Noch-Nicht-Seins* (Frankfurt: Suhrkamp Verlag, 1961), p. 270.

40. Carl Braaten, "Toward a Theology of Hope," *Theology Today* (July 1967), p. 213.

41. See especially in this regard W. D. Marsch, "Hoffen/Worauf?" in *Auseinandersetzung mit Ernst Bloch* (Hamburg: Furche-Verlag, 1963), pp. 91 ff.

42. Leszek Kolakowski, *Der Mensch ohne Altnerative* (München: R. Piper & Co., Verlag, 1960), p. 125.

43. Antoine de Saint-Exupéry, *Terre des Hommes,* quoted in Joseph C. McLelland, "Symbol of Hope for 'Man and his World,'" *The Christian Century,* LXXXIV, No. 28 (July 12, 1967), p. 893.

44. M. M. Thomas and Paul Abrecht, *Christians in the Technical and Social Revolutions of Our Time: World Conference on Church and Society* (Geneva: World Council of Churches, 1967), p. 190. See also Gunter Howe and Heinz Eduard Tödt, *Frieden im Wissenschaftlich-technischen Zeitalter* (Berlin: Kreuz-Verlag, 1966), p. 79.

45. Klaus Tüchel, *Herausforderung der Technik* (Bremen: Carl Schurremann Verlag, 1967), p. 72.

46. Howe and Tödt, *op. cit.,* pp. 38-39.

47. See especially Walter Rauschenbusch, *Christianizing the Social Order* (New York: The Macmillan Co., 1912), and *A Theology of the Social Gospel* (New York: The Macmillan Co., 1917). Also of interest at this point is Benson Y. Landis, *A Rauschenbusch Reader: The Kingdom of God and the Social Gospel* (New York: Harper & Bros., 1957).

48. Landis, *op. cit.,* p. xvi.

49. *Ibid.,* pp. 27-28.

50. Walter Rauschenbusch, quoted in *ibid.,* p. 28 (from *Christianity and the Social Crisis*).

51. Rauschenbusch, *A Theology of the Social Gospel,* quoted in *ibid.,* p. 117.

52. Rauschenbusch, quoted in *ibid.,* p. 18 (from *Christianity and the Social Crisis*).

53. Paul Tillich, *Systematic Theology* (London: James Nisbet & Co., Ltd., 1964), Vol. III, p. 422.

54. *Ibid.,* p. 424.

55. *Ibid.,* p. 274.

56. Thomas and Abrecht, *op. cit.,* pp. 190-191.

57. *Ibid.,* p. 198.

VI

1. Joseph Conrad, quoted in Reuel Denney, *The Astonished Muse* (Chicago: University of Chicago Press, 1957), p. 242.

2. Wiener, *The Human Use of Human Beings, op. cit.,* p. 162.

3. The classic historical studies of this theme are R. H. Tawney, *Religion and the Rise of Capitalism* (New York: Harcourt, Brace and World, Inc., 1926), and Max Weber, *The Protestant Ethic and the Spirit of Capitalism* (New York: Charles Scribner's Sons, 1958); German, "Die Protestantische Ethik und der Geist des Kapitalismus," in *Gesammelte Aufsätze zur Religionssoziologe* (Tübingen: J. C. B. Mohr Verlag, 1947).

4. Margaret Mead, "The Pattern of Leisure in Contemporary American

Culture," in E. Larrabee and R. Meyersohn, *Mass Leisure* (Cambridge: The MIT Press), p. 12.

5. Oscar Cullmann, *Christus und die Zeit: Die Urchristliche Zeit und Geschichtsauffassung* (Zürich: EVZ Verlag, 1962), p. 75.

6. Bertrand Russell, *In Praise of Idleness* (London: George Allen & Unwin Ltd., 1935), p. 2.

7. Harvey Cox, *The Secular City* (London: SCM Press, 1965), p. 173.

8. *Ibid.*, p. 175.

9. See *The International New York Herald Tribune* (May 28, 1968), p. 3.

10. A good exposition of this thesis is the book of *Chicago Sun-Times* reporter William Braden, *The Private Sea: LSD and the Search for God* (New York: Rand McNally Press, 1966).

11. Cox, *The Secular City, op. cit.*, p. 181.

12. "A Soviet Reply to Joint Flight," in *Bulletin of Atomic Scientists,* Vol. XXV, No. 7 (September 1969), p. 15.

13. Helmut Thielicke, *Theologische Ethik II/1* (Tübingen: J. C. B. Mohr [Paul Siebeck], 1959), p. 512.

14. Lilje, *op. cit.*, p. 137.

15. See Werner Linke, *Technik und Bildung* (Heidelberg: Quelle & Meyer Verlag, 1961), pp. 133 ff.

16. Edward T. Hall, *The Silent Language* (New York: Doubleday & Co., 1959), p. 79.

17. Marshall McLuhan, *The Gutenberg Galaxy: The Making of Typographic Man* (London: Routledge & Kegan Paul, 1962), p. 5.

18. Marshall McLuhan, "Sight, Sound and Fury," in Bernard Rosenberg and David Manning White, eds., *Mass Culture: The Popular Arts in America* (Glencoe, N. Y.: The Free Press, 1963), p. 495.

19. Karl Marx, "Privateigentum und Kommunismus," in *Karl Marx–Friedrich Engels Studienausgabe*, Band II (Frankfurt: Fischer Bücherei, 1966), p. 101.

20. Denney, *op. cit.*, p. 242.

21. One need only recall the detached wandering of Lee Harvey Oswald or Richard Speck (who murdered eight nurses in Chicago) or Robert Whitman (who gunned down some forty people at the University of Texas). Mention can be made also of the suspected assassins of Martin Luther King and Robert Kennedy.

22. For a good study on this, see Nels Anderson, *Work and Leisure* (Glencoe, N. Y.: The Free Press, 1961), p. 266.

23. Mead, *op. cit.*, p. 12.

24. Emil Brunner, *The Christian Doctrine of Creation and Redemption* (Philadelphia: The Westminster Press, 1952), p. 66.

25. See at this point the study by Hans-Eckehard Bahr, *Totale Freizeit* (Stuttgart: Kreuz-Verlag, 1963), pp. 61 ff.

26. Frederik Pohl, "The Midas Plague," in *The Case Against Tomorrow* (New York: Ballantine Books, 1957).

27. See the article by David Riesman, "Leisure and Work in Post-Industrial Society," in Larrabee and Meyersohn, *op. cit.*, pp. 363 ff.

28. Helmut Thielicke, *Der Nihilismus* (Pfullingen: Verlag Günther Neske, 1951), p. 48.

29. Ernst Korff, *Betriebs psychologisches Taschenbuch für Vorgesetzte* (Heidelberg: I. H. Sauer-Verlag, 1967), p. 54.

30. Thielicke, *Der Nihilismus, op. cit.*, pp. 200 ff.

31. Hans-Rudolf Müller-Schwefe, *Vom Zuchtvollen Leben: Die Aufgabe der Askese im Technischen Zeitalter* (Hamburg: Furche Verlag, 1959), p. 46 (my translation).

32. Cox, *The Secular City, op. cit.*, p. 173.

33. Korff, *op. cit.*, pp. 50 ff.

34. Bahr, *Totale Freizeit, op. cit.*, p. 41.

Selected Bibliography

A. CYBERNETICS:

Dechert, Charles E. *The Social Impact of Cybernetics.* South Bend, 1966.
Ducrocq, Albert. *Die Entdeckung der Kybernetik.* Frankfurt, 1959.
Frank, Helmar, ed. *Kybernetik: Brücke Zwischen den Wissenschaften.* Frankfurt, 1964.
———. *Kybernetik und Philosophie: Materialien und Grundnuß zu einer Philosophie der Kybernetik.* Berlin, 1966.
Guenther, Gotthard. *Das Bewusstsein der Maschinen: Eine Metaphysik der Kybernetik.* Krefeld and Baden-Baden, 1957.
Halacy, D. S. *CYBORG: Evolution of the Superman.* New York, 1965.
Keidel, Wolf-Dieter, *Kybernetische Systeme des Menschlichen Organismus.* Kön and Opladen, 1963.
Klaus, Georg. *Kybernetik in Philosophischer Sicht.* Berlin, 1961.
———. *Kybernetik und Gesellschaft.* Berlin, 1964.
Linke, Werner. *Technik und Bildung.* Heidelberg, 1961.
Loeser, Franz. *Deontik: Planung und Leitung der Moralischen Entwicklung.* Berlin, 1966.
McLuhan, Marshall. *Understanding Media: The Extensions of Man.* New York, 1964.
Michael, Donald N. *Cybernation: The Silent Conquest* (a report to the Center for the Study of Democratic Institutions). Santa Barbara, Calif., 1962.
Moray, Neville. *Cybernetics.* New York, 1963.
Schneider, Peter Karlfried. *Die Begründung der Wissenschaften durch Philosophie und Kybernetik.* Stuttgart, 1966.
Steinbuch, Karl. *Automat und Mensch: Über Menschliche und Maschinelle Intelligenz.* Berlin, 1961.
———. *Falsch Programmiert.* Stuttgart, 1968.
Strunz, Kurt. *Integrale Anthropologie und Kybernetik.* Heidelberg, 1965.
Tüchel, Klaus. *Herausforderung der Technik.* Bremen, 1967.
Wasmuth, Ewald. *Der Mensch und die Denkmaschine.* Köln and Olten, 1955.
Wiener, Norbert. *Cybernetics or Control and Communication in the Animal and the Machine.* Cambridge, Mass., 1948.
———. *God and Golem, Inc.: A Comment on Certain Points Where Cybernetics Impinges on Religion.* Cambridge, Mass., 1964.

————. *The Human Use of Human Beings: Cybernetics and Society.* Boston, 1954.

B. ETHICS AND THEOLOGY:

Abrecht, Paul, and M. M. Thomas. *Christians in the Technical and Social Revolutions of Our Time.* Geneva, 1967.
Augustine, *The City of God.* New York, 1958.
Bahr, Hans-Eckehard. *Totale Freizeit.* Stuttgart, 1963.
————. *Verkündigung als Information.* Hamburg, 1968.
Benz, Ernst. *Evolution and the Christian Hope.* New York, 1966.
Bloch, Ernst. *Das Prinzip Hoffnung.* Frankfurt, 1959.
Boeck, Hans. *Zur Marxistischen Ethik und Socialistischen Moral.* Berlin, 1958.
Bonhoeffer, Dietrich. *Ethics.* New York, 1955.
————. *Prisoner for God: Letters and Papers from Prison.* New York, 1960.
Chardin, Pierre Teilhard de. *The Phenomenon of Man.* New York, 1959.
Cox, Harvey. *The Secular City.* New York, 1965.
Cruschtschow, N. S. *Über die Kontrollziffern für die Entwicklung der Bolkswirtschaft der UdSSR in den Jahren 1959 bis 1965.* Berlin, 1959.
Gogarten, Freidrich. *Verhängnis und Hoffnung der Neuezeit.* Stuttgart, 1953.
Hatt, Harold. *Cybernetics and the Image of Man.* Nashville, 1968.
Kant, Immanuel. *Critique of Pure Reason.* London, 1934.
Karewa, M. D. *Recht und Moral in der Socialistischen Gesellschaft.* Berlin, 1954.
Kolbanowski, W. N. *Die Kommunistische Moral und die Lebensweise.* Moskau, 1955.
Lehmann, Paul. *Ethics in a Christian Context.* New York, 1963.
————. *Ideology and Incarnation.* Geneva, 1962.
Lohff, H. "Die Anthropologische Fragestellung in den einzelnen Wissenschaften," in A. Filtner, ed., *Wege zur Pädagogischen Anthropologie.* Heidelberg, 1963.
Moltmann, Jürgen. *The Theology of Hope.* New York, 1967.
Müller-Schefe, Hans-Rudolf. *Vom Zuchtvollen Leben: Die Aufgabe der Askese im Technischen Zeitalter.* Hamburg, 1959.
Niebuhr, Reinhold. *The Nature and Destiny of Man.* London, 1941.
Petrossjan, A. A. "Die Ethik und der moderne Rivisionismus," *Philosophische Wissenschaften,* 4, 1958.
Schischkin, A. *Die Grundlagen der Kommunistischen Moral.* Berlin, 1958.
Smith, Ronald Gregor. *The New Man.* New York, 1956.
Thielicke, Helmut. *Der Nihilismus.* Pfullingen, 1951.
————. *The Freedom of the Christian Man.* New York, 1963.
————. *Theologische Ethik I.* Tübingen, 1958.
————. *Wer Darf Leben? Der Arzt als Richter.* Tübingen, 1968.
Tillich, Paul. *Dynamics of Faith.* New York, 1957.
————. *The Protestant Era.* Chicago, 1948.
————. *Systematic Theology,* III. Chicago, 1963.
Wilson, John. *Reason and Morals.* Cambridge, 1961.

C. GENERAL:

Anderson, Nels. *Work and Leisure.* Glencoe, N. Y., 1961.
Arendt, Hannah. *Vita Activa: oder Vom Tätigen Leben.* Stuttgart, 1960.
Baillie, John. *Natural Science and the Spiritual Life.* New York, 1952.
Barbour, Ian G. *Christianity and the Scientist.* New York, 1960.
————. *Issues in Science and Religion.* Englewood Cliffs, N. J., 1966.
Berdyaev, Nickolai. *The Meaning of History.* London, 1950.
Bronowski, J. *The Identity of Man.* London, 1967.
Burtt, E. A. *The Metaphysical Foundations of Modern Science.* New York, 1954.
Butterfield, Herbert. *The Origins of Modern Science.* New York, 1957.
Coulson, C. A. *Science, Technology and the Christian.* Nashville, 1960.
Day, John A. *Science, Change and the Christian.* Nashville, 1965.
Denney, Reuel. *The Astonished Muse.* Chicago, 1957.
Einstein, Albert. *Ideas and Opinions.* New York, 1954.
Ellul, Jacques. *The Technological Society.* New York, 1964.
Farber, Seymour, and Roger Wilson. *Control of the Mind.* New York, 1961.
Fromm, Erich. *The Heart of Man.* New York, 1964.
Hall, Edward T. *The Silent Language.* New York, 1959.
Jonas, Hans. *The Phenomenon of Life: Toward a Philosophical Biology.* New York, 1966.
Korff, Ernst. *Betriebs psychologisches Taschenbuch für Vorgesetzte.* Heidelberg, 1967.
Lilje, Hans. *Das Technische Zeitalter.* Hamburg, 1928.
Lindsay, Jack. *Marxism and Contemporary Science.* London, 1949.
Marcuse, Herbert. *One Dimensional Man.* Boston, 1966.
Matson, Floyd W. *The Broken Image: Man, Science, and Society.* New York, 1964.
McLuhan, Marshall. *The Gutenberg Galaxy: The Making of Typographic Man.* London, 1962.
Michael, Donald N. *The Next Generation.* New York, 1963.
Mumford, Lewis. *Technics and Civilization.* New York, 1934.
Of Men and Machines. Ed. Arthur O. Lewis, Jr. New York, 1963.
Queffelec, Henri. *Technology and Religion.* New York, 1964.
Randall, John Herman, Jr. *The Making of the Modern Mind.* Rev. ed. New York, 1940.
Richardson, Alan. *The Biblical Doctrine of Work.* Naperville, Ill., 1958.
Technology and Christian Culture. Ed. Robert Paul Mohan. Washington, 1960.
Van Leeuwen, Arend Th. *Christianity in World History.* New York, 1946.
Von Bertlanffy, Ludwig. *Problems of Life.* New York, 1960.
Von Weizsäcker, C. F. *The Relevance of Science.* New York, 1964.
Whitehead, A. N. *Science and the Modern World.* New York, 1925.